Disaster Research and Management Series on the Global South

Series Editor
Amita Singh
Centre for the Study of Law and Governance
Jawaharlal Nehru University
New Delhi, India

Disaster Research and Management Series on the Global South is a series coming out of Special Centre for Disaster Research (SCDR) at Jawaharlal Nehru University (JNU), New Delhi, India. SCDR is the first in Asia Pacific to start a course on disaster research within a social science perspective. The series follows and publishes pedagogical and methodological change within the subject. The new direction of teaching, research and training turns from 'hazard based' to 'resilience building'. The series taps such research for the benefit of institutes and higher education bodies of the global south. It also suggests that much of the western literature based upon rescue, relief and rehabilitation which is also being taught in the Asian institutes is not directly relevant to managing disasters in the region. It provides reading and study material for the developing field of disaster research and management.

1. Generates a non-west transdisciplinary literature on disaster research and studies
2. Strengthens disaster governance and improves its legal framework
3. Sensitizes disaster management authorities towards key priorities and attention areas
4. Focus on preparedness is strongly proposed and revisited
5. Highlights changes in pedagogy and methodology of disaster research and teaching
6. Mainstream vulnerable communities of differently abled, elderly, women, children
7. Indicate strategies needed to protect city animals, birds and wildlife during disasters

More information about this series at
http://www.palgrave.com/gp/series/16402

Amita Singh
R. Lalitha. S. Fernando • Nivedita P. Haran
Editors

Development in Coastal Zones and Disaster Management

palgrave
macmillan

Editors
Amita Singh
Center for the Study of Law & Governance
Jawaharlal Nehru University
New Delhi, India

R. Lalitha. S. Fernando
Department of Public Administration
University of Sri Jayewardenepura
Nugegoda, Sri Lanka

Nivedita P. Haran
Additional Chief Secretary, 3rd Floor
Government of Kerala, South Block
Trivandrum, Kerala, India

ISSN 2662-4176 ISSN 2662-4184 (electronic)
Disaster Research and Management Series on the Global South
ISBN 978-981-15-4293-0 ISBN 978-981-15-4294-7 (eBook)
https://doi.org/10.1007/978-981-15-4294-7

This Palgrave Macmillan imprint is published by the registered company Springer Nature Singapore Pte Ltd.
The registered company address is: 152 Beach Road, #21-01/04 Gateway East, Singapore 189721, Singapore

Foreword: A Message from the Coast

Was it time for Mother Earth to take a break? Is this one of the many ways? Has she sent a message to all of 8 billion+? Did she not send messages earlier which we did not heed? Or resisted? Will we learn the lessons? For those fortunate? (Hon'ble Lt. Governor Puducherry, Dr Kiran Bedi on the recent outbreak of COVID-19, March 2020)

Coasts have always been fascinating for me both for their pristine purity of an infinite spiritual universe which they offer and for the challenge which grips me as I stand over a fragile, floating landmass against a mighty, tempestuous and endless spread of turquoise ocean. There is so much to write and feel about the coast which valiantly stands between the two un-equals to protect the enormous biodiversity that adorns her throughout the year. Its ironical that mankind in its miniscule size inhabits the tenuous coast to only weaken it further and make it more vulnerable against the turbulence and fury of the ocean. Coasts need special protection, care and consideration from citizens as much as from governments which control activities over it.

India has a long coastline of 7516.6 km, out of which mainland coastline consists of 6100 km and islands' coastline consists of 1197 km. Around nine states and four union territories fall in the coastal area which establishes that Indian economy has a huge share of these coastal goods, manpower and natural resources. No government can afford to ignore coastal health in terms of natural resources, its culture and its people. My experience of managing disasters at Andaman and Nicobar islands and then as the Lt. Governor at Puducherry has opened me to a deepening understanding about the nature of an ocean and about responsibilities of citizens and administration who inhabit the coast. The manner in which consumerism, ostentatiousness and greed have overtaken mankind has caused the erstwhile serenity of coasts to be somewhat disturbed and disasters to become more frequent, more impactful and more unpredictable. This has only increased as one looks into the scattering of coastal communities and nonhuman wildlife which provided the best early warning system against disasters.

I would prefer to highlight the three most important areas where habitual governmental lethargy and apathy have weakened the coasts; and by mentioning them here, I would like the readers of this work to give a thought to it as they participate in policy planning processes.

First is reinstating participatory governance in coastal cities and villages so that they can protect and conserve their own local resources. I tried to arrange a community meeting at Allankuppam village and the experience was so rewarding for environmental awakening that with the help of these communities I could set up a forum which also included some officials including the area's Junior Engineer. This monthly meeting is now instituted as a regular governance practice. Sometimes, apathetic public servants perceive participatory governance as interference in their day to day affairs because it creates public vigilance and watchfulness. In a recent 11 March judgement in which the Madras High Court dismissed this perception of 'interference' presented as a problem by the Chief Minister against my involvement with communities and local administration revives hope that people are powerful and their rise to vigilance does create ripples across bureaucracy. The judgement has established the need for collaborative, participatory and pro-people governance in the coastal Union Territory of Puducherry. I would like to suggest that this should be the guiding principle of participatory governance on the coasts so that these areas become clean, green and water rich. Notwithstanding the acts and laws through which states and union territories are governed (e.g., Puducherry Village and Commune Panchayats Act of 1973), there is a pressing need for some serious incorporations and changes in the Disaster Management Act 2005, which has so far not attempted to elaborate the responsibilities of decision makers too intricately with a vision for sustainable participation and administrative accountability in managing a disaster.

Second is the building of true leadership which can keep the coasts prepared to prevent a hazard from escalating into a disaster. This may entail a responsibility of keeping all ministries coordinated and not just the State Disaster Management Authority. That would be over-expectation from a single department for action which takes place in other departments. I have been a great advocate of building a grassroots women leadership which brings immense resilience to local community action. I even instituted a Puduvai Pengal Award to encourage women leaders of local communities. The first awardee of this honour is Mrs Allimuthu of Villianur who maintains 51 women Self-Help Groups, and has spearheaded the construction of 100 toilets I also suggest that police force at the rank of Sub-Inspectors should be trained to be more community oriented to prevent land grabbing in coastal areas by powerful interests.

Third, that governance should be 'conservation oriented' – conservation of everything which is local such as culture, people, livelihood choices, plants, trees, animals, birds, mountains, hills and agricultural practices. A combination of conservation knowledge which is indigenous wisdom and modern technology would definitely help in sustaining coastal wetlands for all forms of life that inhabits it. Conservation also has a spiritual element which I experienced while walking by an 'old reborn tree' collectively conserved near the police station at Puducherry(see the photograph).Such old trees have been preserved through titles of 'sacred groves' or the 'abode of deities'. In reality, they are the greatest insurance for the local communities against scarcity of water, livelihood, food and ecosystem. The Forest Rights Act of 2006 has given community rights to manage these pristine environmental conservation spaces. Those belonging to the coasts are the biggest stakeholders in conservation but plans undertaken by coastal governments have not trusted their participation and sharing of knowledge. The energy of college youth and children in schools ought to be tapped to achieve a two sided goal of sustainable development and sustainable economy at the coasts. It had been a worthwhile experience to organize college level Science Exhibitions and open to them a world of relevant and appropriate scientific solutions for their surroundings. Many bigger institutions of Science in the neighbourhood such as the National Centre for Sustainable Coastal Management at Chennai should play a more involved community role to tap young talent of local students. An experience of conservation of water bodies and tanks has generated much responsibility. When Puducherry youth undertook desilting of Bahour Lake, much support arrived from Club Mahindra and Sri Aurobindo Society for undertaking the work. One sixth standard girl promised that she and her friends would take care of it. The Aurobindo Society has decided to adopt her to be an engineer. Another experience of recharging local people was the Clean Kanagan Lake. However, the most promising effort was the revival of a huge water body next to the Puducherry Medical College which was dead under the loads of garbage dumps and sewerage discharge of drains. Its revival into a tourist spot now is a result of a combined and collaborative effort of many government departments, local communities, Resident Welfare Associations, children and college students. This was part of a Green Puducherry Initiative.

In conclusion, I found that four steps definitely lead to achievement; Plan purposefully, Prepare prayerfully, Proceed positively and Pursue persistently. I applied these principles during my directives to disaster management departments at Puducherry. I also found that the Karaikal administration had undertaken a meticulous preparation and planning as a result of which no life was lost. In encountering disasters at the coasts, mankind is merely a dot before the might and prowess of the ocean. What it can do is to change ways to live and use coastal attributes and resources. An attitude of conservation and co-existence is indispensable. I wish to recall the passion of Dr Jane M.Goodall, an English primatologist and anthropologist. At the age of 26, carrying just her notebook and binoculars, she travelled from England to an unknown tropical forest of Tanzania to protect, save and conserve Chimpanzees and to give the world a remarkable window into humankind's closest relatives. What if she had not dared to do this? Chimpanzees would have gone extinct and with them a deep history of the evolution of human beings. It is passion which drives exceptional leaders and the coasts need these leaders to train mankind to face oceans, save planktons and enrich carbon sequestration to obviate climate change and disasters waiting to devour our sinking coasts.

> *What you do makes a difference and you have to decide what kind of difference you want to make, The greatest danger to our future is apathy* (Dr Jane M. Goodall)

(This is an invited 'Foreword' from Dr Kiran Bedi, a legendary Indian Police Service (IPS) Officer, currently the Lt. Governor of the Union Territory of Puducherry. She is one of the most brilliant and dedicated citizen of mother earth, a Magsaysay Award winner, a governance reformer, the Asian Tennis Champion and a versatile grassroots social worker heading NGOs 'Navjyoti' and IV Foundation. She has courageously withstood anyone who misused power against a common citizen in every way. Her contributions to coastal Puducherry have brought a transformation in many ways. Coasts have always needed someone like her to decide on how to protect them)

Lieutenant Governor of Puducherry
Puducherry, India

Dr. Kiran Bedi, IPS (Rtd)

Acknowledgments

Our deepest gratitude to the former Executive Director of the National Institute of Disaster Management (NIDM) Mr. Anil Kumar and to the current Executive Director Major General Manoj Kumar Bindal for enabling wholesome participation, providing an atmosphere of academic freedom and encouraging empirical research in the NIDM faculty. The faculty of the Special Centre for Disaster Research (SCDR) at JNU and the NIDM collectively encountered challenges in the field and contributed to a variety of curriculum building events at JNU.

SCDR is equally appreciative of Lt. Gen. N.C. Marwah, Member of the National Disaster Management Authority (NDMA), who consistently provided the required backup support during field surveys and data collection in disaster-affected areas. Sometimes the victims of disasters became his mission and brightened studies on administrative leadership in managing disasters.

The JNU Central Library deserves a special mention. The readiness to help in exploring and providing rare literature and data on disasters came in handy to many of the authors. The former librarian Dr Ramesh C. Gaur and after him Dr Manorama Tripathy understood and appreciated our search for literature on transdisciplinary disaster studies in law, governance, public policy and sciences. We will never forget the vibrancy and intellectual energy of a younger librarian, Azmi Khan, of a smaller library at the Centre for the Study of Law and Governance at JNU. She always waited to hear from the research team about their next search and joined them in loud youthful discourses.

Our office staff of the Special Centre for Disaster Research, Deepak Kumar, Hemchand Pandey, Darakshan, Jhuman and Narinder, brightened the work with their secretarial assistance and coordination support, which helped us immensely in meeting the deadline.

This section also acknowledges the unforgettably versatile contribution of the research team of Gaurika Chugh, Vualzhong Mung, Chetana Attri and Natasha Goyal. The long-distance support of Dr Manika Kamthan and Dr Binod Kumar was always available like disciplined soldiers of the team.

The most enlightened support arrived from the Springer-Palgrave publishing team guided by the visionary Editorial Director for Springer Singapore William Achauer, Associate Editor for Palgrave Macmillan Sandeep Kaur and Editor for Springer-Nature Nupoor Singh, who managed discussions and coordination in an unfailing manner. Their collaboration goes beyond mere publication, as they absorbed themselves in the research teams including many workshops, symposiums and debates.

Publishing these volumes of non-western literature has led the SCDR research team to look for many new authors from local administrations, affected communities and implementers. The editors acknowledge the shared contribution of many who, despite the motivation, could not write due to their intensive work responsibilities in the Chennai floods, Cyclone Gaja and the Kolkata Bridge collapse. The Indian Council of Social Science Research (ICSSR) has empowered many of these implementers who remain knowledge repositories for original literature in disaster studies. The editors appreciate the supportive role of ICSSR in bringing about this volume.

Last but not the least, NAPSIPAG (Network of Asia Pacific Schools and Institutes of Public Administration and Governance) stands with this initiative, as strong and determined as ever before. This is one big Asia-Pacific family of policy experts which is always passionate to celebrate collaborations in generating knowledge from their homelands.

Contents

List of Figures

Loss and Damages from Cyclone: A Case Study from Odisha, a Coastal State

Artificial Intelligence Based Early Warning System for Coastal Disasters

List of Tables

Disaster Management in Coastal Areas: An Introduction

Nivedita P. Haran

Introduction

The coastal regions are five times more vulnerable to disasters than the hinterlands. About 80% of disasters are linked to water and the coastal areas are most exposed to water. More than 80% of the earth's surface is water and now humans are in search of water on the moon. The majority of countries in Asia have are near the sea. India, a sub-continent in itself, has a 10,000 km coastline. Sri Lanka, an island country, is situated at the confluence of two seas and an ocean. Bangladesh is considered one of the most vulnerable countries in terms of the vagaries of disasters that are connected to the sea; its geo-morphological features are such that a large part of its coastline lies at a level lower than the Marine Spatial Planning (MSP).

The coastal areas are exposed to a number of disasters that emanate from water: cyclones, storms, floods, high tidal waves and tsunamis. Due to easy access to water coastal areas are more densely populated. River-mouths are also the location for some of the larger urban agglomerations:

N. P. Haran (✉)
Government of Kerala, South Block, Trivandrum, India

A. Singh et al. (eds.), *Development in Coastal Zones and Disaster Management*, Disaster Research and Management Series on the Global South, https://doi.org/10.1007/978-981-15-4294-7_1

ports attract trade that leads to human settlement. Recent times have seen some of the most severe urban floods in these cities. Urban flood brings with it destruction of life and property, homelessness, damage to infrastructure and break-down of public civil service. Urban floods bring life to a standstill causing huge loss in commercial activities. It is estimated that each day Mumbai suffers floods causes a loss of over $ 300 million to its economy. In the aftermath of floods, issues such as the lack of potable water, outbreak of water-borne diseases and work days lost are alarming. Surely no country can afford such loss.

The coastal zone also provides some of the most precious and priceless resources: marine and aquatic life, rare metals mined from the sand, vegetation and fruits, thorium from the sparkling sands in Kerala, and gourmet seafood.

Nature is a generous giver. But when it is mindlessly and selfishly exploited, nature's fury can also be so overpowering that the impact can be unendurable. Humans are the only species that seem to grab and destroy as if there will be no consequences. It is ironic that humans are identified as occupying the apex of the evolutionary pyramid in terms of rationality, as the damage done to coastal areas in most developing countries has reached a stage where it is now almost irreparable. True, poverty can be a reason behind the destruction of coastal areas: mangroves uprooted for firewood, unlimited harvesting of marine products, fish, seaweed and even coral, and encroachment into the sea. Hence 'Eradication of poverty', as SDG 1 reminds us, remains the primary responsibility and the goal of any government, and indeed any society. The issue that needs to be raised here is the systematic destruction of the coastal areas through governmental mismanagement and human greed. In order to protect the long coastline along the Indian subcontinent a number of legislations are in progress: regulating construction, harvesting of marine products and use of marine transport. Yet in blatant disregard for and violation of these statutes and their poor enforcement, commercial construction has been allowed. The Coastal Regulation Zone norms, for instance, lay down strict guidelines on the nature of permissible construction and at what distance from the coastline. But the norms have been amended and made more lax each time, allowing more and more construction. Waste from these construction projects is released into the sea. A visit to any beach exposes one to a pathetic sight: solid waste including non-biodegradable waste is seen lining the beach area or floating as flotsam.

Areas that are declared as protected as ecologically sensitive zones or biologically sensitive or as a protected site based on the Ramsar declaration are not left unexposed. Land has always been the most acutely impacted asset that has suffered at the hands of the unholy nexus between crooked realtors, unscrupulous politicians and mendacious public servants. It is this unholy trinity that forms the basis for crony capitalism. So, even though the blame is placed on the poor and vulnerable, it is the rich and privileged that form the bulk of the land-grabbers. The coastal areas fall within their radar for three reasons: the coastline anywhere is geographically one of the most scenic by far, with the presence of water, vegetation and salubrious climate. Second, due to the above reason these lands are also the most high-priced and their location is considered the most upmarket. Third, most of the coastal lands are of low gradient, and hence easier to construct. They are also easily reachable by good access routes. Taking advantage of these factors, construction in violation of norms is clearly visible: posh hotels, high-end residences, even highly-rated and coveted commercial premises.

The disregard shown towards the Western Ghats, for instance, that run parallel to the west coast of India is a case in point. The report of the expert committee that recommended caution was ignored. Whether there is a causal link between the systematic destruction of the ghats and the severe climatological changes that have occurred in this area in the last few years could be the subject of debate and further research. But the fact that such destruction is irreversible is undeniable. The frequency of high rainfall incidents in Mumbai has increased. The extraordinarily intense rainfall over the state of Kerala in August 2019 is another case in point. The coastal regions are extremely sensitive and tampering with their natural features is indeed very risky.

Yet, mankind does not seem to learn these lessons. Cases of building collapse consequent to heavy rainfall and floods are reported with alarming frequency. Some of these constructions are either buildings declared unsafe or are constructed in violation of coastal zone or building norms. There was the case in Kochi, Kerala where a realtor had constructed cottages and even high-rises destroying virgin forests off the coastline. Of course, such extensive construction could not happen without the connivance of the administration. Based on a public interest litigation, the court had registered a prima facie case, but such cases get dragged on for years, if not decades. Citizens need to take matters into their hands. If clients refuse to buy properties that are constructed in violation of the building by-laws, that ignore environmental considerations, and that destroy the ecology and pollute the environment, such agencies, corporations, authorities and even officials will be socially ostracized. Punitive action should be prompt and effective to act as a glaring deterrent so that the economic loss is also clear and accountability fixed unequivocally.

It is high time that the policy-makers and law-makers are made conscious of the concept of 'public trust'. The coastal areas and water bodies are part of public assets that are to be protected for future generations. It is the duty of public servants to hold public assets in public trust and to protect, conserve and maintain them. Unfortunately, over the years 'public trust' has been diluted and public assets are either ignored or treated as bureaucrats' or the elected representatives' personal assets. As a result coastal land is allowed to be grabbed haphazardly and construction is carried out with no consideration for the environment or ecology, and all to bestow favors and illegal gratification.

The chapters selected for this anthology have been researched and written by erudite scholars. The scholar's or academic's duty does not end with the publication of a paper; it goes far beyond it. Scholars hold the intellectual authority that can argue and establish a hypothesis through cogent debate and logical deduction. It is their duty to present and disseminate their theses to the citizens who are also the stakeholders. If they are truly committed and passionate about their ideas they need to present them to those who can make a difference: the policy-makers. There is a need to bridge the gap between thought and action, cognition and operation. In a perfect world such a gap should not exist, but in most countries it does, to a lesser or greater extent. It is as much the duty of academics as of administrators to close this gap to ensure that there is a synergetic outcome to improve the world.

The coastal regions of South Asia have faced some devastating disasters: the tsunami of 2004 left the southern coast of India and many parts of Sri Lanka reeling. Cyclones of varying intensity hit coastal areas of India and Bangladesh with unerring regularity. Storms and high tidal waves can be frighteningly destructive; when occurring in unison heavy rainfall and high tides have brought Mumbai to a standstill. Water bodies can be unforgiving agents of destruction. The movement of the sea leads to coastal erosion. Interference with the water in one area through construction and landfills can have ramifications for another area quite unknown to technical experts.

In this age of information explosion it is important to collate relevant data for use in research and policy-making and to make it available to those concerned. With the help of AI, data analysis is child's play. Vulnerability maps of the coastline can help identify the areas prone to erosion and accretion, which would provide basic data for development activities: road alignment, bridge location, areas requiring vegetative protection, or engineering interventions through sea-walls, gabion boxes or groins. Engineering interventions can be used sparingly and only where unavoidable. With reliable and accurate maps that are geo-referenced, it is possible to manage coastal disasters more effectively. There are also systems in place to monitor tremors that occur due to shifts in tectonic plates under water, tremors that can result in tsunamis; such tremors are now being monitored round the clock. With the use of technology the occurrence and arrival of storm surges can be forecast with precision.

The impact of climate change can be significant in the coastal areas. In some countries there are special statutes to manage coastal-linked disasters. Perhaps it is time for a discussion in this direction to commence in the South Asian countries. After all, adaptation to climate change is also very important in the coastal areas. The impact could range from rise in sea-level, thus inundating coastal areas, to disappearance of marine species and alteration in mangroves and other vegetation. Similarly, specially trained forces are needed to face such disasters. The example set by the state of Odisha in facing cyclones and the systemic changes made has borne fruits: the loss in human life and property has decreased considerably. There is a lesson in this for other areas similarly placed. It is possible to be better prepared to face coastal disasters with dedication. There is no reason why with modern technology and communication systems human life and all life and property should be placed in danger.

Nivedita P. Haran Additional Chief Secretary (Rtd), Government of Kerala. Led the constitution of the first State Disaster Management Authority after the Disaster Management Act 2005. Currently based in Delhi

PART I

Policies, Law and Regulations for the Mitigation of Coastal Disasters

Coastal Conservation in Sri Lanka: Problems and Prospects

R. Lalitha S. Fernando, H. M. I. U. P. Herath, and R. B. P. M. Rathnayake

Human Onslaught Over the Coasts

A coastal zone is considered as the seaside terrain and it is the interface between the mainland and the sea. Coasts are unique, valuable and often threatened areas where the sea meets the land (Goodwin 2015). Coastal zones are continually changing due to the dynamic interaction between the oceans and the land (Nelson 2018).

Sri Lanka is an island enriched with various kinds of coastal resources which consist of a 1620 km long coastline all around the country (Lakmali et al. 2016). Due to rapid increase of population and developmental projects near coasts, these areas have particularly become more vulnerable. Due to various development activities in relation to various sectors such as construction, engineering, tourism, fishery and cultivation, coasts are degraded; some are able to be mitigated while some are not.

R. L. S. Fernando (✉) • H. M. I. U. P. Herath • R. B. P. M. Rathnayake
Department of Public Administration, University of Sri Jayewardenepura, Nugegoda, Sri Lanka
e-mail: rlsf@sjp.ac.lk; isharaherath@sjp.ac.lk

A. Singh et al. (eds.), *Development in Coastal Zones and Disaster Management*, Disaster Research and Management Series on the Global South, https://doi.org/10.1007/978-981-15-4294-7_2

Coastal line in Sri Lanka is nowadays being threatened by different causes including increasing population pressure. The natural environment is being converted into artificial ports, tourist beaches and residential places. The result is severe erosion of beaches and excessive sedimentation (United Nations Environment Programme n.d.). Rising sea levels, topological disasters, soil erosion by water and degradation of the natural seascape are identified as broader problems which may influence coasts (Encyclopedia 2018).

Sri Lanka has a nearly 1700 km long coastline and a 30,000 km^2 continental shelf area up to 120 m deep (Koralagama 2008). As Prasada et al. (2015) cited from the Ministry of Fisheries and Aquatic Resources Development, the coastal zone contains 26 major fisheries harbors; 58 boat anchorages; 193 improved landing centres; 890 minor fish landing centres; a fishing fleet of 51,127 boats of different sizes; a marine fishing household population of 824,680; and five seaports. The above statistical data shows the nature of activities in the relevant area and the extent of responsibility coming under the scope of marine environment protection in the country. Thus, maintaining coastal environment protection becomes paramount.

Population growth threatens the coastal zone (Samaranayake 1997). Once the population starts to increase, resources fall short with respect to population growth, therefore overutilization, overexploitation and overextraction is justified for survival. Coastal erosion has become a severe problem that results in damaging or destroying buildings, coastal structures and other infrastructure of the country. It also causes loss or degradation of valuable land and disrupts fishing, navigation, recreation and other activities (Coast Conservation Department 1997). A coastal departure, due to erosion by the sea, caused the loss of several square miles of the coast, especially in the southwestern region in Sri Lanka (Amarasinghe and Gerritsen 1976). Prasada et al. (2015) noted that threats are still prevalent in the sustainability of a healthy marine environment in Sri Lanka. The National Report of Sri Lanka has identified many threats to the coastal and marine environment and its living resources, coastal and marine habitats, shoreline stability, coastal and marine fisheries, brackish water fisheries and culture, coastal and marine biodiversity (Prasada et al. 2015).

Tsunami was a huge threat that badly impacted the coasts. The shoreline in Sri Lanka was severely affected and eroded with debris and seawater which resulted in destroying cultivatable lands, paddy fields and natural vegetation in the area (Nayanananda 2007). The tsunami in 2004 effected Galle district severely by destroying 70% of buildings located on the coastline and nearly 30% of structures within 1 km inland (UN Office for the Coordination of Humanitarian Affairs 2004).

Releasing garbage into the sea and coasts is another dangerous problem which has massively threatened natural environment and living beings in the sea and related areas too. There are frequent reports on the occurrences of dumping of ship-generated waste in the ocean, causing serious environmental and economic damages (Marine Environment Protection Authority 2014). Sunday Times (2016) reported that Sri Lanka has become a reported country in garbage dumping into the sea. Sri Lanka 'Global Coastal Index' represented a very bad picture with the island being ranked fifth out of 20 countries identified for dumping polythene and plastic to the ocean (Daily News 2017). Tissera (2018) pointed once that the coastal waste becomes more disruptive after the rains, weekends and during festivals. Sri Lanka Coast Guard (SLCG) works closely with the Maritime Environment Protection Authority in marine cleaning operations and also in terms of solid coastal litter; the situation is worse in the Southern and the Western Provinces.

Storms can impact the coastline at a higher rate than others. As per Lakmali et al. (2016), in the North-east coastline severe long-term erosions can be observed in some places like Verugal, and after the starting of the Pulmudei plant, the sediment supply has been reduced to downstream and thus it has resulted in downstream erosion.

Coral and sand mining in the coasts is another problem that effects coastal erosion. Improper mining for coral, sands, seashells, and limestone may adversely affect the coastal zone and the sustainability of coastal habitat. Lakmali et al. (2016) noted that earlier, many locations in these coastlines were subjected to severe erosions, for example Lansigama, Uswetakeiyawa, and and so on, and this was mainly due to the use of sea sand as an alternative to river sand policy.

"Recreations" done around beaches are another well-known cause for coastal erosion. In Sri Lanka, this has become prevalent and can be seen as beach-oriented resort development that has taken place in Negombo and

Bentota on the west coast, Kalkudah, the North of Trincomalee and Arugam Bay on the East coast, and Hikkaduwa on the South coast (Sri Lanka Tourism Development Authority 2019). The coastline of the Western province is the most visited tourist destination in Sri Lanka and many of these tourists travel a long distance to observe nature while the same experience can be provided within the basin with less travel time and this would also contribute to the enhancement of local economy (Ministry of Agriculture 2017). Therefore, it is impossible to expect the end of these activities and hence, the coastal areas will be prone to gradual degradation.

In recent times, the erosion problem has been heightened by consequences of sand and coral mining and other engineering and industrial activities in the coastal area by numerous parties and the rapid development of the tourism industry in the last decade has also had a huge impact on this matter. Under the country's development plans, priority is undoubtedly given to the preservation and environmental control of the beaches and for the effective management of the coasts. Thus, the development activities need to be in tandem with these conservation plans too.

This chapter examines the coastal management planning and its effectiveness in Sri Lanka with reference to the local context based on the secondary data.

What Is a Coastal Zone?

Coastal zone can be identified as the 'dividing boundary' between sea and land (Isobe 1991). Coastal regions are intensely dynamic areas and of critical importance to humans. Coastal conservation act of Sri Lanka (1981) defines Coastal Zone as the area lying within a limit of 300 m landwards of the Mean High Water line and a limit of 2 km seawards of the Mean Low Waterline. Coastal zones contain a variety of Earth's most complex and diverse ecological systems, productive both biologically and economically.

Coastal erosion weakens the coastal zone despite being a natural process that occurs whenever earth/sand is eroded from the shoreline. It is commonly referred to as the loss of landmass into sea due to natural processes such as waves, winds, and tides, or even due to human interference (Science Daily 2019). At present, coastal erosion is very common and about 70% of the sandy coasts are subjected to erosion around the world (YeYincan 2017).

Forty percent of the world's population lives not far from the coast and rely on coastal and marine ecosystems, habitats and resources for food, building materials, building sites and agricultural and recreational areas, while utilizing coastal areas as a dumping ground for sewage and garbage too (Ducrotoy 2019). The sea margins are affected almost everywhere by man, and intrusion on coastal areas continues worldwide (Schubel 1994). Human actions impact massively on coastal erosion; they cannot be limited. Causes for coastal erosion can be categorized as: natural causes, sedimentation and human activities (World Register of Introduced Marine Species 2019).

According to the World Bank, human actions on coasts and sea have destroyed 20% of mangroves, 30% of seagrass beds and 20% of coral reefs. The total length of the world's coastlines is approximately 504,000 km (World Ocean Network 2013). This mass area helps thousands of living beings to find food and to engage in their industries worldwide. From the total of world's population, 66% lives within 150 km of the coast (Goodwin 2015). This high population density near the coast is a major reason for causing coastal degradation, which results from a higher level of utilization. Erosions due to regular waves of monsoons cause long-term hazard as it is uncontrollable. Due to human-induced greenhouse effect, natural and other man-made causes, sea level is rising continuously. It is expected that within the next 300 years, sea level will rise by a further 5 meters (World Ocean Review 2010). Natural causes result in the washing away of 9.3 billion tons of soil a year, but human intervention increases that to almost 24 billion tons (UIA 1998).

Due to coastal erosion, coastal land is being swallowed by the seawater and coastal villages and houses are forced to move inland, losing the living places of humans, destroying beach biodiversity and ecological balance, and harming human lives and natural environment (YeYincan 2017). Thus, coastal erosion seems to change from a natural environment change to a serious hazard. This serious and offensive issue can be seen in Sri Lanka. With the impact of the tsunami in 2004, it was deeply subjected to erosion and, hence, various efforts are required to conserve the coast.

Rationale for Conservation

'Coast conservation' means the preservation and protection of the coasts from erosion or by the sea, and also includes the planning and management of development actions within the Coasts (Coast Conservation Act No 57 1981). Disasters are impossible to prevent; suitable strategies should be followed in order to reduce the impact of disasters to coasts. There are so many advantages of conserving the coastal line. Nayanananda (2007) noted that the coastal area has become a major part of the economy that contributes to the national GDP, which is more than 40% of the total. With industries like fisheries, tourism and other mineral resources, the economic contribution of the coastal sector cannot be ignored. Many people's livelihoods are basically tied in with food production and coastal fishing communities (Nayanananda 2007). Tourism in coastal areas is also economically important. Visiting the white sandy beaches and coral reefs are some of the most favorite activities of both locals and foreigners. These activities provide employment for local people (International Union for Conservation of Nature 2014). Ultimately, these income-generating activities are beneficial for communities.

Conservation efforts are challenged due to the following eight factors out of which many factors are beyond the control of one country alone. They are:

1. Sea level rising is another factor of coastal erosion. Due to climate change and global warming, average sea levels have increased over 8 inches (about 23 cm) over a century ago, and every year, the sea rises by another 0.13 inches (3.2 mm) (Nunez 2019).
2. Tsunami is a huge threat that badly impacts the coasts. Predictions of tsunami flood levels and potential effects of tsunami flooding are generally known as the greatest risk for coastal areas (Camfield 1994). It is a giant sea wave generated by earthquakes, volcanic eruptions, or landslides and such waves can reach up to 30 m in height and have potential to wipe out coastal cities (Nelson 2018). When tsunami waves become high, they severely affect coastlines, causing property damage and loss of lives (SMS Tsunami Warning 2018). Damage from a tsunami occur to structures located at the shoreline or along river channels near the shoreline and bridges or scouring of the channel bottom near bridge piers (Camfield 1994).

Among other causes, the tsunami is the most dangerous coastal disaster that causes deep damage to coastal areas.

3. Coastal landslides are another kind of natural disaster which may cause serious coastal damages. Historically, coastal landslides have produced substantial damage to urban and tourist settlements, port facilities and road and railway networks and people were made homeless (Iadanza et al. n.d.).
4. Great storms can cause erosion at a higher rate. During storms, beaches can erode rapidly and heavy waves can cause rapid damage along the coast (Nelson 2018). This massive wave energy often causes beaches to sacrifice large amounts of coastline and this may result in a permanent loss of land.
5. Sedimentation due to man-made activity may strongly influence the tendency of coastlands to be eroded. For example, construction of coastal structures (such as seawalls) leads to changes in coastal transport pathways, resulting in erosion in some areas and accretion in others (Geoscience Australia n.d.).
6. Coastal construction and industries have become a major cause of coastal degradation in the long term. As many important industries and centers are built near urban areas and ports, coasts are increasingly vulnerable to damage. Also, main industrial activities include iron melting and processing, chemical and petrochemical industry, paper mills, vehicle factories, shipbuilding, power plants and food processing (including fish), which affect coastal areas (Ducrotoy 2019). Construction activities often cause permanent destruction or decrease of coastal habitats, due to land assertion, extraction of bottom material, dumping and disposal of waste.
7. Dredging or sand mining can also be associated with accidental erosion. For economic and other purposes, coastal resources are subjected to be mined inappropriately (Prasetya n.d.). At larger scales, natural and human-induced climate change can modulate the rate and probability of coastal erosion. Among the adverse effects of actions by men, construction of ill-designed harbors and coastal protection works and the mining of sand and coral from various coastal areas are significant (Amarasinghe and Gerritsen 1976). Excessive mining for coral, sands, seashells and another coastal habitat adversely affect the coastal region and its sustainability too (Geoscience Australia n.d.).

8. Dredging and dumping in the sea are also bad practices followed worldwide. Many developing counties and poor nations tend to release all garbage nearby or into the deep sea. According to the Ocean Health Index (2019), marine waste, including plastics, wood, paper, metal and other manufactured materials is found on beaches and ocean worldwide. About 60–80% of all marine debris is plastic and Ocean Conservancy's Trash Free Seas Alliance states that more than 8 million metric tons of plastic are dumped into the ocean each year (Pawar and Shirgaonkar 2016). It seems that coasts are eroded and will also diminish as a result of waves hitting weakening shoreline.

Integrated Coastal Zone Management (ICZM)

Isobe (1991) suggested a Theory of Integrated Coastal Zone Management (ICZM) which consists of three functions: disaster prevention, utilization, and ecology. The author revealed that in the existing Coastal Zone Management System, there is no mechanism of integrated coastal zone managing mechanism. The basis of ICZM is that it must endeavor to maintain a healthy environmental foundation. Further, the ICZM model emphasizes that the status of the environment determines the status of coastal zones. The theory always highlights the importance of maintaining a good environment where its people could live in a better protective environment. The objective of ICZM is to achieve a new state of equilibrium without causing destructive instability. It is expected to maintain the health and safety of the environmental foundation which contributes to three areas: disaster prevention, utilization, and ecology. A smooth flow among water, soil and energy is needed to achieve this end and through this process, the environmental capacity of the coastal zone will be increased, and humans will realize the consequent benefits (Isobe 1991).

Japan's first formal coastal zone management scheme was embodied in the Coastal Act of 1953 and its objective was to prevent disasters. But it was not developed from an integrated viewpoint. There had been Japanese laws such as the Fishing Port Act, Harbor Act and the Public Waterfront Landfill Act set down separately which covered the conservation and utilization of Japan's coastal zone and those different laws resulted in differing responsibilities by different government agencies (Isobe 1991).

The ICZM emphasizes disaster mitigation. As mitigation is important to coastal zone management, the safeguarding of the environmental foundation is exceptionally important. Artificial changes in the coastal zone must be done in such a way that they do not destabilize or adversely affect the environmental infrastructure, for example, water quality, soil quality, or air quality (Isobe 1991).

There are some practical examples that can be seen in ICZM in Japan where the Dashihira Dam in Fujiyama was built so that sand accumulated on the upstream side of the dam is exorcized through a sand flush gate, this ensuring a supply of soil downstream (Isobe 1991). The maintenance of coasts provides another example of ICZM that runs with a long-term broad-based perspective of coastal management and thus a healthy and a stable environmental foundation is the main goal of ICZM (Isobe 1991). It seems that Japan has a well-managed coastal environment which is hazard protected.

Empirical Research on Coastal Conservation

Cristie et al. noted that Marine Protected Area (MPA) is the ideal management tool for Marine Ecosystem-Based Management (EBM) efforts and the success of MPAs is positively correlated with remoteness, low population and the threat from external forces such as illegal fishing. Similar findings were confirmed by Pollnac (2001) and Maypa (2002). Secondly, Cristie et al. noted that developing leadership and the technical skill levels of NGOs and municipal government is a good investment. Thus, all the parties related to local governments should be more knowledgeable and skillful in managing coasts. As the third enforcement, fisheries management regulations, is an ongoing challenge everywhere, when legitimate rules are disarrayed, punishments should be given in strict and fair manner. As a fourth factor, Patrick (2009) describes that in order to have opportunities for new ideas through participatory planning processes, conflict-oriented scenarios are needed to be encouraged that can lead to success at the end.

Addo and Addo (2016) carried out a study adopting a scientific approach to estimate the rate of erosion and tested the results against perceived erosion trend by the local settlers. It was revealed that about 79% of the coastline is eroding, while the remaining 21% is stabilized against each causal factor. The study highly emphasizes indigenous knowledge of localities on coastal erosion in coastal communities; the level of that knowledge differed based on their geographical location. According to the results, the local knowledge system, which the indigenous people gain over time through experience, highly impact coast management in a particular area. Further, Addo et al. in the same study, highlighted the essence of improving the local knowledge that should be integrated with scientific data to achieve sustainable management of the coast. Local knowledge can provide cursory information on the coastal erosion situation and the scientific data can provide the information statistically regarding the impact of coastal erosion on livelihoods that can be collected through interviews with the local community.

Siry (2007) introduced the Decentralized Coastal Zone Management system for Southeast Asia. The concept of decentralization lies in the notion of a certain transfer of power from the upper level to lower governance (Siry 2007). Further, the author explains the necessity of an Integrated Coastal Management (ICM) mechanism as the coastal zone will remain productive only if there is an integrated or a holistic approach. The coastal zone enables management while maintaining environmental, social and economic needs together. As a management approach, ICM covers all aspects of the coasts, such as economic activities, planned development, natural resource utilization and conservation and addressing multi-stakeholders and conflicts (Siry 2007). Decentralized CZM is the combination of decentralization and ICM approaches. Siry (2007) emphasized that this concept intends to maximize the performance of delegated responsibilities for managing the coasts. His results found that decentralized CZM work requires a careful consideration of at least three factors, understanding and awareness on decentralized policy for managing coastal resources; mechanisms for upward and downward accountability; and mechanisms for assigning responsibilities, authorities and resources.

Best Practices of Coastal Conservation

This section provides a brief discussion of the best practices that were undertaken worldwide related to coastal conservation. Some of these practices are applicable to the local context to some extent.

Country Initiatives in Coastal Management

Integrated Management of Marine and Coastal Resources: Australia

Cummins et al. (2004) discusses many important initiatives based on the integrated management of coastal resources that ensured conservation and sustainability of the coastal zone in Australia. This study showed that Australia has given a high level of priority to conserve marine biodiversity and also proactively considered a broad range of economic, social and cultural aspirations.

Mitigation of Landslide Risk: Italy

Iadanza (n.d.) noted that, in order to mitigate the landslide risk of coasts in Italy, several corrective actions have been taken: slope redesign and rock scaling; passive structures like rockfall barriers, wire mesh nets, rockfall tunnels; retention structures including walls, piles, reinforced soil embankments, gabions; dowels and anchors; soil reinforcement and erosion control measures; subsurface and deep drainage; protection measures at the foot of cliffs against sea wave action (e.g. breakwaters).

Mangrove Belt: Malaysia

Othman (1991) noted that nearly 30% of the coastline in Malaysia is undergoing erosion; because mangroves are known to reduce wave energy as waves travel through them, the Department of Irrigation and Drainage in Malaysia has imposed a rule that at least 200 meters of mangrove belts must be kept between the bunds and the sea to protect the bunds from eroding. This is important for all countries to follow as it is beneficial for the natural environment too.

Hard and Soft Engineering: Indonesia

According to Prasetya (n.d.), in Indonesia, a combination of hard engineering approaches (breakwaters/jetties/revetments) of different shapes are attached to functional design and aesthetic values, and soft engineering approaches (beach nourishment) are also used in certain ways. As per the author, they succeeded in stopping coastal erosion on most of the coasts.

Coastal Zone Management Programs: The United Kingdom

The United Kingdom does not have any overall legislation for national coastal zone management. Coastal management, therefore, falls under the control of certain statutory and non-statutory instruments which guide individual sectors rather than geographical areas (Cummins et al. 2004). As per the author, the procedure employed in the United Kingdom is basically top-down but recognizes voluntary initiatives whereby local voluntary media work within the bounds of national, sectorally driven legislation, towards the goal of achieving sustainable coastal management.

Integrated Coastal Zone Management Plan (ICZM) in the Netherlands

The Netherlands also adapted integrated coastal zone management plan (ICZM). Their goal was not only to reinforce the coastal flood but also to increase the quality of the region, and to that end, everyone has become responsible, including national government. The government imparts overall guidancethrough local institutes by working with civil society, residents and businesses (Veersalu 2011).

Hard Engineering and Soft Engineering

According to IDAC (2007), coastal protection can be ensured through natural or manmade systems or an amalgamation of both. Available techniques for coast conservation fall into two categories: hard strategies and soft strategies. A well-considered coastal engineering solution can provide the structural/hard and non-structural/soft protection where beneficial effects are maximized while minimizing adverse effects of potential solutions to problematic areas on a coast.

Hard engineering or structural protection is done to maintain the coastline's integrity, where rigid or semi-rigid structures are constructed along the coastline to battle deformation from wave or disruptive actions and structures are built in areas that face erosion (IDAC 2007). For example, seawalls are designed to resist extreme storm events, prevent excessive overtopping and preserve existing infrastructure, by completely blocking the waves and their effects. Life span of a sea wall is approximately 75 years and can reduce the erosion of the beach and also socially comforting for local inhabitants (S-Cool n.d.).

Barrier islands are another important form of structural protection, particularly along the Atlantic coast, south of Long Island and these islands help to protect the mainland from heavy waves from the sea; however, many have been so highly developed that their own coasts may need protection as well (Pile Buck Magazine 2018).

Seawalls are other large coast-conserving structures built using different types of construction materials like rubble or concrete. There are many examples in Sydney estuaries where mangroves have been established or have reestablished naturally in front of seawalls, in some cases among the gaps between the rocks making up the seawall (Catchment Management Authority—New South Wales Government 2009).

Soft engineering implies that the natural environment is used in reducing coastal erosion and river flooding (Tutor2u 2019). It does not involve building artificial structures but utilizes a more natural and a sustainable approach. Soft coastal protection can be the result of the following diverse strategies: dunes, supratidal beaches, foreshore, mega and morphological nourishment (IDAC 2007). In beach replenishment, which is usually done once a year using trucks, is very costly and time-consuming (S-Cool n.d.). Over the next 12 months, the material is washed away by shore drift before being replaced and the final method of coastal management is, of course, to do nothing and let the sea to attack the coastline naturally (S-Cool n.d.).

In 'Beach nourishment', sand is pumped onto a beach to build it up; since the sand gets combined with the existing beach, the beach becomes larger and larger beaches appeal to tourists. This method is more practical. Lincshore is the largest beach nourishment scheme in the UK, covering beaches from Mablethorpe to Skegness (Royal HaskoningDHV 2019). As another soft-engineering technique, reprofiling means redistribution from the lower part of the beach to the upper of the beach (BBC 2019). Williams (2005) defines that it is an artificial reshaping of beaches using existing beach materials. Further, he shows the examples of beach reprofiling on the South-East England coast undertaken by the Environmental Agency.

Afforestation is another method of soft engineering where trees are planted in a drainage basin to intrude and store rainwater, thereby slowing the movement of rainfall towards the river and then reducing the amount of water discharged (Tutor2u 2019). Flood plain zoning is another soft-engineering technique and, it prohibits building on flood plains so that the river can flood naturally and the cost of flooding is reduced, causing less or no damage to property (Tutor2u 2019).

These soft- and hard-engineering techniques could be applied by considering the severity of the coastal vulnerability and the nature of coastal erosion.

Coastal Conservation Efforts in Sri Lanka

As an island, Sri Lanka possesses various coastal resources. A coastal protection unit was established in Colombo port commission in 1963. In 1981, the parliament enacted Coast Conservation Act, No. 57 of 1981, and accordingly the coast conservation division was upgraded to Coast Conservation Department (CCD), which has been vested with a director (Parliament of Sri Lanka, 2011). An output of CCD is an environmental-friendly development, positive influence on the national economy, secure value of coastal community, efficiency and productivity including mitigation of disaster impacts on residents of coastal areas (Coast Conservation Department 2019).

An advisory council was established to review coastal management problems where instructions were given by the Minister in charge of coastal conservation (Samaranayake 1995). Under the act No. 57 of 1981 a coastal zone management plan has been established. The plan deals with problems of erosion, losses and degradation in coastal areas. According to the regulations of coastal conservation, coral resources cannot be removed for anything rather than research purposes and mining of sand. Further, archeological sites and any other development activity which has a significant influence to degrade the quality of natural zones are prohibited to be established within 200 meters by the department of coast conservation. (Samaranayake 1995).

After amending the act as the Coast Conservation (Amendment) Act, No. 49 of 2011, it extracted several conditions in coastal conservation as follows: 'make provision for a survey of the coastal zone and the preparation of a coastal zone management plan, regulate and control development activities within the coastal zone, make provision for the formulation and execution work for coast conservation and resource management within the coastal zone, make consequential amendments to certain written laws and provide for matters connected therewith or incidental thereto' (Parliament, 2011, Part 5).

Departments and Agencification of Coastal Conservation in Sri Lanka

Coastal management has been in operation in Sri Lanka from the 1920s, yet the coastal conservation division was established as late as 1978. It was upgraded to the coastal conservation department under the ministry of fisheries in 1984 (Silva et al. 2011). Still, coastal conservation department remains as the primary agency in charge of coastal zone management in Sri Lanka. However, there are several other agencies that are responsible for conserving the coastal regions.

The department is headed by the director general of coastal conservation and there are several divisions: coastal development division, research and design division, monitoring and evaluation division, coastal resource management, finance, administration and legal unit (Coastal Conservation Department 2019).

Specifically, the objectives of the coast conservation department are to enhance the existing prestige of coastal environment, maintain and develop the shoreline, promote economic development through the coastal resources and improve the lifestyle of coastal communities (Coastal Conservation Department 2019). The services which are provided by CCD are shoreline monitoring, formulation and designing shoreline stabilization, handling emergency coastal erosion, constructing coast protective structures, formulating and implementing strategies, plans, policies and projects to restore the ecosystem and conducting awareness programs for the public.

The CCD has a responsibility to regulate development activities by associating with setback standards, environmental impact assessment and monitoring compliances, updating and expanding the coastal risk on erosion, awareness of tsunami and other disasters and establish and maintaining coastal green belt through community participation (Coastal Conservation Department 2019).

There are several institutes that are responsible for conserving the coastal areas, such as department of wildlife conservation, which operates in the protected areas including marine protected areas, and ministry of fisheries and ocean resources, which manages coastal and marine biodiversity and habitats. Marine pollution prevention authority addresses the problems of marine pollution and is highly concerned about hazardous waste disposal techniques and prevention of pollution from ships. National aquaculture development authority is focused on the challenges faced by the coastal zone and recovery of the damages (Silva et al. 2011). Several institutes in Sri Lanka practice various physical planning and regulating systems for the development activities for coastal conservation. Hence, national physical planning department has introduced the coastal zone management plan in 2004 to develop Coastal Fragile areas. Central environment authority is a regulating authority to control pollution in the environment which cause adverse impacts on coastal zones.

Similarly, the land reclamation and development corporation in Sri Lanka has recognized the significance of coastal resource management. Moreover, Sri Lankan tourist board provides tourist infrastructure and activities of coastal zone and further developments. Urban development authorities and municipal councils and other councils have the authority to enhance development activities and protect the environment, including coastal zones (Silva et al. 2011).

Plans and Projects Formulated for Coastal Conservation in Sri Lanka

The economic significance in coastal areas is emphasized by Coastal Zone Management (CZM) plan which commenced in 1990 and is reviewed about once in four years (Nayanananda 2007). This plan emphasizes to develop, use and conserve the resources in coastal areas by integrating coastal zone management in a sustainable manner. The goal of establishing the CZM plan is to identify the issues and problems that are required to be addressed in the coastal zones and the plan contributes to managing coastal resources sustainably.

The plan further explains the necessity of such a plan as natural process is threatened in the coastal areas increasingly by storm surges, which cause coastal erosion, and human activities that lead to erosion and pollution. The CCD is the main agent who is responsible for employing strategies to reduce or mitigate the worst environmental problems affecting Sri Lanka's coastal areas. The CCD issues permit for all kinds of development activities near coasts. It includes residential and commercial constructions, recreational structures, roads, sewage treatment plants, dredging, grading, sand mining and breaching of sand bars.

'Coastal 2000' is a more important concept in coastal regions which provides economic significance and the objective of this concept is to formulate strategies and programs for the second generation of coast management (Nayanananda 2007). The Costal 2000 addresses primary issues in the coastal management programs and it includes coastal resource management by governmental and non-governmental institutes. This is a collaboration of Coast Conservation Department (CCD), National Aquatic Resource Development Agency (NARDA), Central Environmental Authority (CEA), Irrigation Department (ID), Ministry of Fisheries and Aquatic Resources (MFAR) and other agencies (Nayanananda 2007).

Furthermore, the department of coast conservation and coastal resource management has conducted several projects in order to conserve the coastal regions. Beach nourishment projects, coastal hazard and risk assessments, rehabilitation of revetments, off-shore sand mining and converting of existing groins to an off-shore breakwater are the completed projects initiated by the department. Construction of coast protection scheme, a research and information Centre at Nilaweli and a water quality management programme are the ongoing projects (Coastal Conservation Department 2019). Although the department sources have given several titles for projects, there are effects of booming population growth, economic and technological development that threaten the coastal ecosystem and resources.

Samaranayake (1995) noted that setback is a healthy practice where it leaves a minimum sea level line; it has been defined as an area that is left free from physical modifications. Setback exemptions are decided by advisory council of coast conservation for several development actions. Further, the author noted that the Environmental Impact Assessment (EIA) is a parallel theory which is a kind of a system that is concerned to have a significant influence on the coastal environment and engages in development activities within the coastal zones (Samaranayake 1995).

Green Borders ('Haritha Theera'), which is an effective plan implemented by the CCD under the guidance of director general is a successful preservation procedure to prevent coastal erosion. Other than that, this is a good solution to transfer unstable coastal areas to stable areas. For example, a Green border was proposed to be established in Unawatuna coastal area to enhance its quality. Green borders were planned by the participation of the provincial councils, divisional councils, non-governmental organizations, schools and universities to make the concept effective (Rathnayake 2015).

Strengths of Coastal Conservation in Sri Lanka

As a strategy for coast conservation in coastal regions in Sri Lanka, 'Coastal 2000', that provided a direction to coastal resource management, was prepared in 1992 and is currently being implemented (Coastal Conservation Department 2019). The concept has strengthened to manage proper coastal ecosystems through the current institutional and human capacity which may overcome the problems related to the coastal damages. As the key sectors of the coastal region, fishing, tourism, trade and manufacturing have provided comparative advantages. Especially, in Sri Lanka, the most conventional economic activities in coastal cities are coastal tourism and marine fishing. Through better conservation tactics, coastal areas have attained several development goals so far. The study of the economic significance of coastal region of Sri Lanka (2007) showed a higher level of economic development in coastal regions during the last two decades through tourism, fishing industries and services, which are the main sectors in national development.

As the major foreign exchange earner, coastal tourism generates foreign exchange to help development. Therefore, the conservation of coasts is a major component to achieve that level. Even though Sri Lanka is facing challenges like tremendous poverty, disparity among regions and slower acceleration of economic growth, coastal strategies could meet those objectives because a majority of the population live in coastal regions (Nayanananda 2007).

Under the CZM plan, Sri Lanka has adopted several management strategies, regulations, awareness programmes, education, planning, monitoring, researches and policy development practices. As an example, Hikkaduwa and Rekawa had a special area management plan which had a beneficial outcome associated with CZM plan. Through the experience of executing the CZM plan, it has been identified that participatory approaches are necessary to manage Sri Lankan coastal habitats (Coast Conservation Department 1997).

In the study of Unawatuna coastal area (2015), it was found that a programme was conducted by local council with the support of the department of coast conservation and coastal resource management to prevent the removal of waste products into the sea by hotels and other related agencies and to maintain a clean environment. Also, the coastal conservation authority and Habaraduwa police station exhibit notice boards about the preservation of the coastal area to educate the tourists (Rathnayake 2015). This strategy contributed to the progress of handling coastal resources effectively.

Further, the department has stopped illegal constructions, which are mentioned in coast conservation act to preserve the coastal areas. Under the guidance of the coastal conservation department, two main processes of conservation of the coastal areas are planned and designed as follows: making one beach entrance for every 500 m by removing buildings and walls that are constructed across the beach and creating green borders along the beach (Balasooriya 2010 cited by Rathnayake 2015). Well-designed and well-operating strategies and plans will reap benefits and overall development. Further, the government should pay attention and sponsor such schemes to conserve the coastal areas.

Drawbacks of Coastal Conservation in Sri Lanka

Coastal Zone Management Plan in 2004 was formulated to be implemented throughout the coastal zones; however, due to the absence of a definition of a Coastal Fragile Area (CFA), the plans were unclear and it has still not come into effect (Silva et al. 2011). The study further noted that the mission of Coast Conservation Department for the management of the coast and the jurisdictions to carry out the mandates have certain mismatches.

Whether there are comprehensive policies for coastal management, the real challenge at present is the inability of coastal conservation department to coordinate diverse policies and legal mandates in coastal zones. Further, the department regulates activities which affect the coastal zones, for example, environment impact assessment that was conducted in Muthurajawela sand fill (Silva et al. 2011). As weaknesses of the implementation of coastal conservation management, limited development projects have been called under the environmental impact assessment, for example, coal-fired thermal power plant projects, hotel constructions and fishing harbours (Samaranayake 1995). As a good tactic, environmental impact assessment can be used to enhance the capacity of conserving coast resources sustainably. Therefore, relevant parties should give their courtesy to wider spread of those strategies into almost all development projects.

To be accomplished, the pre-planned activities need to have management approaches at all levels of administration and adequate human and financial capacities. Therefore, maintaining and implementing such approaches every year require a greater budgetary allocation (Silva et al. 2011). Most of the training development and awareness programmes are lacking in practical aspect of financial capacity and trained human resource. Since lowest income and lower level educated levels live in coastal cities, they are not willing to give up their livelihood activities like fisheries and they have no idea regarding sustainability of their occupation which is a major issue (Silva et al. 2011).

Coastal pollution is a global phenomenon that arises from incorrect inland aquatic ecosystems because of the complexity of coastal habitats and the magnitude of human association (Arunashantha 2015). Further, the author identified several issues in the Sri Lankan context such as lack of reliable information, lack of sufficient awareness on conservation and sustainable resource usage, insufficient participation of community in decision making and resource management, several environmental degradations taking place in coastal areas, poor law enforcement, lack of proactive techniques used during emergencies and lack of integrated management approaches. Further, Silva et al. (2011) noted that almost all the services in Sri Lanka have been politicized; therefore, a higher political interference for the bureaucracy, which has lost its independence, is a major threat for public services including conservation of coastal resources.

Even though Sri Lanka has several regulations related to coastal protection, the legal and penal actions are not applicable to coastal resource management. For example, poor coastal communities are not in a position to provide information and do not submit applications to grant permits to build houses because poor coastal communities consider this regulation procedure as a barrier to their livelihoods (Arunashantha 2015).

Senevirathne noted that main coastal issues are indefensible resource utilization and absence of adequate planning and management initiatives which have created serious environmental problems in coastal zones. The author further explained that sustainable management and scheduled efforts must focus on special area management to achieve the desired goals and objectives of sustainable coastal management in Sri Lanka.

Suggestions for Improving Coastal Conservation in Sri Lanka

During the past two decades, coastal management requires various management techniques and need more than one agency. The narrow definition of 'the coastal zone' has given an insufficient basis for the effective management of shorefront erosion and construction. Coastal zone regulation needs to be revised and each responsible party, such as local-level and provincial-level officers and coastal communities should be involved in decision making and formulating strategies (Samaranayake 1995). Thus, the laws, regulations and other mandates should be reinforced to have a possible impact on coastal conservation, otherwise, the mismatch will be long-lasting.

To be successful in well-designed coastal resource management, Sri Lanka should enhance the capacity of relevant authorities and individuals by awarding and doing certain short-term and long-term training programmes and by promoting coastal resource management-related national policies (White 2006). Kirupakaran and Thiruchelvam have given suggestions to uphill community contribution through awareness programmes and it will implement conservation efforts to preserve the uniqueness of biodiversity which has potential in developing nature of tourism in Sri Lanka.

Australia provides an example by ensuring better implementation of coastal conservation programmes at each level and has attempted to give a higher priority to marine biodiversity, which is recognized as a fundamental factor of coastal conservation. Sri Lanka could adapt such meaningful programmes which had the best results in advance.

Enhancing the awareness of coastal communities can minimize the conflict among development and coastal resources that leads to the prevention of the damage to the coastal resources (Nayanananda 2007). Fishing is the traditional livelihood of the majority of the coastal community and it will remain the same in the future. To protect sustainability, measures should be indicated to increase the profit level. Through coast conservation, ultimately it is expected to achieve sustainability which focuses to meet the requirements of the present without bargaining the ability to meet the needs of the next generation.

Sri Lanka should devise long-term efforts to conserve coastal areas like Belgium which has a wider communication capability with a higher range of audience; it has also adopted financial allocations to a long-term practice.

It is necessary to develop coastal environment, legal institutes and socio-economic practices, formulate draft management plans, consult local people, implement pilot projects and feasibility studies and train personnel for the projects (White 2006).

In the Italian context, a lesson for a meaningful solution to erosion is rockfall barriers. They have a secure funding arrangement for ICZM and have designed a strong management at the national, regional and local level. Therefore, every level plays a significant role and valuable lessons have to be learned through their experiences (Veersalu 2011).

Post and Lundin (1996) presented numerous principles and characteristics in integrated coastal management, namely, tending to be more sectorially oriented by moving beyond the traditional methods, recognizing the connectivity between coastal system and to have a multidisciplinary and holistic perspective, promoting awareness among government and residents about the significance of the concept of sustainability and environmental protection and handling the balance of protective ecosystems and coast-dependent economies. These principles have a similar match with the weaknesses of Sri Lankan coastal zones.

As a best-practicing country, the Netherlands has accepted to use Integrated Coastal Zone Management plan (ICZM) and they are highly recommended to be incorporated among authorities, stakeholders and general public for the coastal conservation (Veersalu 2011). Hence, collaborative participation is a vital. Enhancement of coast conservation act and increase public awareness on coastal resources is a must (Alagan 2009).

Tourism, fishing, trade and industries are key sectors in coastal regions which make a vast impact on economic development. Sri Lanka needs to be developed further in its capacity of financial, human and other managerial aspects in order to attain the level of success. Instead of that, to achieve a broader aspiration, Sri Lanka should have a healthy vision, well-developed objectives, certain strategies, projects and programmes. That will speed up the progress of developing coastal conservation in a productive way.

Conclusion

The study has identified several problems faced in coastal areas and the causes of mismanagement of coastal conservation. The coastal area in Sri Lanka should be conserved and prevent its negative influences on the biological and ecological balance of the area through respective authorities. To overcome issues generated in coastal regions, best practices used by other countries can be adopted such as in Japan, Italy, Malaysia and Indonesia, which have a larger extent of coastal areas. In the study, the theories related to coastal conservation in Sri Lanka and worldwide have been discussed in an attempt to identify irrelevant practices that can be implemented in Sri Lanka. Further, existing acts, laws, main structure of coast conservation and coastal resource development and other related institutes have been analysed to identify their individual responsibility.

Sri Lanka has followed various strategies, plans, projects in implementing effective conservation of coastal resources and those tactics have generated strengths to build and increase public awareness on coastal resources (Alagan 2009).

Sri Lanka has a comprehensive and better coast conservation plan. However, in practice, several issues and weaknesses could be identified which decline the level of coastal conservation. To overcome the shortages, several suggestions are proposed.

Limitations

This study is mainly based on secondary data, which is the main limitation of this study. A comprehensive analysis with empirical data could be undertaken in future research. As the subject is a more significant area in Sri Lanka, further research is needed with primary data.

References

Addo, K., & Addo, I. (2016). Coastal Erosion Management in Accra: Combining Local Knowledge and Empirical Research. *Journal of Disaster Risk Studies*. https://doi.org/10.4102/jamba.v8i1.274.

Alagan, R. (2009). *Sri Lanka's Environmental Challanges*. www.researchgate.net.

Amarasinghe, S., & Gerritsen, F. (1976). *Coastal Problems in Sri Lanka*. Retrieved June 3, 2019.

Arunashantha, H. (2015). Over Utilization of Coastal Resources and Its Impact: The Case of Sri Lanka. *Social Investigation*, pp. 175–185.
BBC. (2019). *Coastal Management.* Retrieved June 14, 2019, from BBC https://www.bbc.com/bitesize/guides/z2234j6/revision/2
Camfield, F. (1994). Tsunami Effects on Coastal Structures. *Journal of Coastal Research*, 177–187. Retrieved September 16, 2020, from http://www.jstor.org/stable/25735597
Catchment Management Authority – New South Wales Government. (2009, June). *Environmentally Friendly Seawalls.* Sidney: Department of Environment and Climate Change NSW. Retrieved from https://www.hornsby.nsw.gov.au/__data/assets/pdf_file/0005/107528/Environmentally-Friendly-Seawalls.pdf
Coast Conservation Act No 57. (1981). *Coast Conservation Act No 57.* The Parliament of the Democratic Socialist Republic of Sri Lanka.
Coast Conservation and Coastal Resource Management Department. (2019, February 27). *About Us: Department of Coast Conservation.* Retrieved from Department of Coast Conservation Website http://www.coastal.gov.lk/index.php?option=com_content&view=frontpage&Itemid=1&lang=en
Coast Conservation Department. (1997). *Coastal Management Plan, Sri Lanka.* Coast Conservation Department. Government of Sri Lanka, Colombo.
Cummins, V., Mahony, C., & Connolly, N. (2004). *Review of Integrated Coastal Zone Management & Principles of Best Practice.* Coastal and Marine Resources Centre. Retrieved from https://www.ucc.ie/research/crc/papers/ICZM_Report.pdf
Daily News. (2017, November 17). *Marine Pollution*, Sri Lanka. Retrieved from http://www.dailynews.lk/2017/11/17/features/134764/marine-pollution
Dissanayake, U. (2005). *Strategies for the Improved Management of Coastal Zone of Sri Lanka.* University of Moratuwa, Department of Civil Engineering. Retrieved from http://dl.lib.mrt.ac.lk/bitstream/handle/123/1389/pre-text.pdf?sequence=4&isAllowed=y
Ducrotoy, J. P. (2019). *Threats to the Coastal Zone.* Retrieved from http://marine-species.org: http://www.coastalwiki.org/wiki/Threats_to_the_coastal_zone
Encyclopedia. (2018, June 11). *Coastal Erosion.* Retrieved from UIA http://encyclopedia.uia.org/en/problem/132952
Geoscience Australia. (n.d.). *Coastal Erosion.* Retrieved June 17, 2019, from Australian Government Geoscience Australia https://www.ga.gov.au/scientific-topics/hazards/coastalerosion
Goodwin, B. (2015, January 11). *Why Are Coasts Important?* Retrieved June 14, 2019, from prezi.com: https://prezi.com/jhjbzfochjhs/why-are-coasts-important/

Iadanza, C., Trigila, A., Vittori, E., & Serva, L. (n.d.). *Landslides in Coastal Areas of Italy*. Retrieved June 4, 2019, from https://sp.lyellcollection.org/content/specpubgsl/322/1/121.full.pdf

IDAC. (2007). *Coastal Protection*. IDAC. Retrieved from https://www.iadc-dredging.com/ul/cms/fck-uploaded/documents/PDF%20Facts%20About/facts-about-coastal-protection.pdf

International Union for Conservation of Nature. (2014, August 13). *Protecting and Sustaining Human Life: Why Healthy Coastal Ecosystems Are So Important*. Retrieved from https://www.iucn.org/content/protecting-and-sustaining-human-life-why-healthy-coastal-ecosystems-are-so-important

Isobe, M. (1991, July 1). A Theory of Integrated Coastal Zone Management in Japan. *ESENA*. Retrieved from http://www.glocom.ac.jp/column/1991/07/a_theory_of_integrated_coastal.html

Koralagama, D. (2008). Community Perception Towards a Set Back Area: A Case Study in Galle District, Sri Lanka. *IIFET*. Vietnam.

Lakmali, E., Deshapriya, W., Jayawardene, K., Raviranga, R., Ratnayake, N., Premasiri, H., & Senanayake, I. (2016, April). Long Term Coastal Erosion and Shoreline Positions of Sri Lanka. *Journal of Survey in Fisheries Sciences*. Retrieved from http://sifisheriessciences.com/article-1-90-fa.pdf

Marine Environment Protection Authority. (2014). *Home*. Retrieved from Marine Environment Protection Authority http://www.mepa.gov.lk/web/index.php?option=com_content&view=article&id=46&Itemid=34&lang=en

Maypa, A. P. (2002). Long-Term Trends in Yield and Catch Rates of the Coral Reef Fishery at Apo Island Central Philippines. *Marine and Freshwater Research*, 207–213.

Ministry of Agriculture, L. I. (2017). *Western Provincial Biodiversity Profile and Concervation Action Plan*. Ministry of Agriculture, Land, Irrigation, Fisheries, Animal Production and Health, and Agrarian Development. Govt. of Sri Lanka, Ministry of Agriculture Battaramulla.

Nayanananda, O. (2007). *The Study of Economic Significance of Coastal Region of Sri Lanka in the Context of Environmental Changes of Pre and Post Tsunami*. Colombo: Coast Conservation Department and The Ministry of Environment and Natural Resources.

Nelson, S. (2018). *Coastal Zones*. Tulane University, Dept. Earth & Environmental Sciences. Spring. Retrieved from https://www.tulane.edu/~sanelson/Natural_Disasters/coastalzones.htm

Nunez, C. (2019, February 19). *Sea Level Rise, Explained*. Retrieved from National Geographic https://www.nationalgeographic.com/environment/global-warming/sea-level-rise/

Ocean Health Index. (2019). *Trash Pollution*. Retrieved from Ocean Health Index http://www.oceanhealthindex.org/methodology/components/trash-pollution

Othman, M. (1991). Value of Mangroves in Coastal Protection. Paper avaialble at https://water.gov.my/jps/resources/auto%20download%20images/5844e2d2c84ce.pdf, accessed August 20, 2020
Patrick Christie, R. B. (2009, April 17). Back to Basics: An Empirical Study Demonstrating the Importance of Local-Level Dynamics for the Success of Tropical Marine Ecosystem-Based Management. *Coastal Management*. https://doi.org/10.1080/08920750902851740.
Pawar, P., & Shirgaonkar, S. (2016). *Plastic Marine Debris: Sources, Distribution and Impacts on Coastal and Ocean Biodiversity.* Retrieved from https://www.researchgate.net/publication/295919494_Plastic_marine_debris_Sources_distribution_and_impacts_on_coastal_and_ocean_biodiversity
Pile Buck Magazine. (2018, March 27). *Basics of Coastal Engineering*. Pile Buck International. Retrieved from https://www.pilebuck.com/marine/basics-coastal-engineering/
Pollnac, R. B., Crawford, B., & Gorospe, L. G. (2001). Discovering Factors that Influence the Success of Community-Based Marine Protected Areas in the Visayas, Philippines. *Ocean & Coastal Management, 44*(11–12), 683–710. https://doi.org/10.1016/S0964-5691(01)00075-8.
Post, J., & Lundin, C. (1996). *Guidelines for Integrated Coastal Zone Management* (Environmentally Sustainable Development studies and Monographs Series No. 9). Washington, DC: The World Bank.
Prasada, S., Fernando, R., Arachchige, B., & Konasinghe, K. (2015). Complexities in Individual Capacity in Interagency Collaboration: The Case of Law Enforcement in Marine Environment Protection in Sri Lanka. In *International Conference on Business Management*. Colombo.
Prasetya, G. (n.d.). *Thematic Paper: The Role of Coastal Forests and Trees in Protecting Again Coastal Erosion*. Retrieved June 10, 2019, from fao.org: http://www.fao.org/3/ag127e/AG127E09.htm
Rathnayake, M. V. R. M. S. (2015). Negative Environmental Impacts of Tourism in Unawatuna Beach Area. Samaja Vimarshana, Faculty of Social Sciences, University of Kelaniya, 01: 139–152.
Raymundo, L., & Maypa, A. (2002). Recovery of the Apo Island Marine Reserve, Philippines, 2 Years After the El Niño Bleaching Event. *Coral Reefs, 21*, 260–261. https://doi.org/10.1007/s00338-002-0237-6.
Royal HaskoningDHV. (2019). *Lincshore – Beach Renourishment*. Retrieved from royalhaskoningdhv.com: https://www.royalhaskoningdhv.com/en-gb/united-kingdom/projects/lincshore-beach-renourishment/822
Rupasinghe, M., & Perera, S. (2014). *Costal Ecosystems: Hazards, Management and Rehabilitation*. NAM S&T Centre.

Samaranayake, R. (1995). *Resources Volume for Tertiary Level Education and Training in Coastal Zone Management*. Colombo: Coast Conservation Department.

Samaranayake, R. (1997). *Sri Lanka's Agenda for Coastal Zone*. ICG Publishing. Retrieved June 3, 2019.

Schubel, J. (1994). *Coastal Pollution and Waste Management*. The State University of New York. Retrieved June 17, 2019, from https://www.nap.edu/read/2249/chapter/10

Science Daily. (2019). *Coastal Erosion*. Retrieved from Science Daily https://www.sciencedaily.com/terms/coastal_erosion.htm

S-Cool. (n.d.). *Coastal Management Strategies*. Retrieved June 3, 2019, from S-Cool https://www.s-cool.co.uk/gcse/geography/coasts/revise-it/coastal-management-strategies

Silva, S. d., Sellamuttu, S. S., Kodituwakku, D. C., & Atapattu, S. (2011). *Governance Performance in Integrated Coastal Management*. Colombo: IUCN (International Union for Conservation of Nature) Sri Lanka Country Office.

Siry, H. (2007). *Making Decentralized Coastal Zone Management Work for the Southeast Asian Region: Comparative Perspectives*. New York: Division for Ocean Affairs and the Law of the Sea Office of Legal Affairs, the United Nations New York. Retrieved from https://www.un.org/depts/los/nippon/unnff_programme_home/fellows_pages/fellows_papers/siry_0607_indonesia.pdf

SMS Tsunami Warning. (2018). *Tsunamis: The Effects*. Retrieved from SMS Tsunami Warning https://www.sms-tsunami-warning.com/pages/tsunami-effects#.XPUHMYgzbIU

Sri Lanka Tourism Development Authority. (2019). *Projects at SLTDA*. Retrieved from Sri Lanka Tourism Development Authority https://www.sltda.lk/projects

Sunday Times. (2016, May 1). *News: Sri Lanka Among the 'Dirty Five'*. Retrieved from The Sunday Times http://www.sundaytimes.lk/160501/news/sri-lanka-among-the-dirty-five-191570.html

Tissera, H. R. (2018, February 11). *Sea of Trash: Inland and Overseas Garbage Washes Up on Lanka's Beaches*. Sri Lanka. Retrieved from http://www.sundaytimes.lk/180211/news/sea-of-trash-inland-and-overseas-garbage-washes-up-on-lankas-beaches-281482.html

Tutor2u. (2019). *Soft Engineering*. Retrieved June 14, 2019, from tutor2u.net: https://www.tutor2u.net/geography/reference/soft-engineering

UIA. (1998, May 4). *Soil Erosion by Water*. Retrieved from Encyclopedia http://encyclopedia.uia.org/en/problem/132939

UN Office for the Coordination of Humanitarian Affairs. (2004). *Tsunami Disaster – Sri Lanka Summary Situation Report 26–31 Dec 2004*. reliefweb. Retrieved from https://reliefweb.int/report/sri-lanka/tsunami-disaster-sri-lanka-summary-situation-report-26-31-dec-2004

United Nations Environment Programme. (n.d.). *Coastal Zone Management*. Retrieved June 3, 2019, from United Nations Environment Programme https://www.unenvironment.org/explore-topics/oceans-seas/what-we-do/working-regional-seas/coastal-zone-management

Veersalu, T. (2011). *The Best Practices of Coastal Zone Protection and Conservation in Spatial Planning*. Estonia: Estonian University of Life Sciences Institute of Agricultural and Environmental Sciences.

Williams, R. (2005). *Beach Recharge in Sussex & East Kent: A Preliminary Inventory & Overview*. University of Sussex. Retrieved from http://www.sussex.ac.uk/geography/researchprojects/BAR/publish/Phase-1-final-Beach%20recharge.pdf

World Ocean Network. (2013). *Coastal Management – Facts and Figures*. Retrieved from World Ocean Network https://www.worldoceannetwork.org/won-part-6/carem-wod-2014-4/thematic-resources-coastal-management/facts-figures-coastal-management/

World Ocean Review. (2010). *Living with the Oceans. A Report on the State of the World's Oceans*. Retrieved from World Ocean Review https://worldoceanreview.com/en/wor-1/coasts/sea-level-rise/

World Register of Introduced Marine Species. (2019). *Natural Causes of Coastal Erosion*. Retrieved from World Register of Introduced Marine Species http://marinespecies.org/introduced/wiki/Natural_causes_of_coastal_erosion

YeYincan. (2017). *Coastal Erosion*. Elsevier. https://doi.org/10.1016/B978-0-12-812726-1.00007-3.

R. Lalitha S. Fernando serves as a senior professor and chair of the Department of Public Administration, Faculty of Management Studies and Commerce of the University of Sri Jayewardenepura in Sri Lanka. She also serves as the director of the Research Centre for Governance and Public Policy of the university. She served as the former head of the department of Public Administration and also the chairman of Research Committee of the Faculty of Management Studies and Commerce of the university and secretary general of the Network of Asia- Pacific Schools and Institutes of Public Administration and Governance. She has published a number of research papers, book chapters, and articles related to public management and environmental management at both the national and international levels. She is a co-editor of several books on disaster management.

H. M. I. U. P. Herath serves as an assistant lecturer in the Department of Public Administration, Faculty of Management Studies and Commerce, and the University of Sri Jayewardenepura, Sri Lanka. Previously, she served as a research assistant attached to the Research Centre for Governance and Public Policy in the university. She completed her BSc Public Management (Special) Degree with a First Class from the University of Sri Jayewardenepura, Sri Lanka.

R. B. P. M. Rathnayake serves as an assistant lecturer in the Department of Public Administration, Faculty of Management Studies and Commerce of the University of Sri Jayewardenepura, Sri Lanka. Previously, she served as a research assistant attached to Research Centre for Governance and Public Policy at the University. She completed her BSc Public Management (Special) Degree with a First Class from the University of Sri Jayewardenepura, Sri Lanka.

The Coastal Zone Policy in Bangladesh: An Appraisal

N. Nabila Hoque

Introduction

The coastal zone of Bangladesh covers nineteen southern districts with a total land area of 47,201 square kilometres. Bangladesh's coastal region covers about 20% of the nation's total land area and more than 30% of the country's cultivable lands. The Bay of Bengal is the largest among 64 bays in the world and an estimated 1.4 billion people live along its coastline in Bangladesh, India, Thailand and Myanmar. The coastal area of Bangladesh is full of natural resources, it consists of an extremely versatile ecosystem, for example, the world's largest single tract of mangroves (the Sundarbans), the world's longest beach (the Cox's Bazar), coral reefs, dunes, wetlands and so on. Nevertheless, the said region is often the victim of Mother Nature's wrath in the form of disasters such as tropical storms, storm surge and heavy and frequent downpour. The life and livelihood of the people in the coastal zone are also adversely affected and disrupted by various other naturally occurring and human-induced disasters, that is, water and soil

N. N. Hoque (✉)
British School of Law, Dhaka, Bangladesh

A. Singh et al. (eds.), *Development in Coastal Zones and Disaster Management*, Disaster Research and Management Series on the Global South, https://doi.org/10.1007/978-981-15-4294-7_3

salinity, numerous forms of pollution, land erosion, elevated arsenic content in land and water, climate change risks, and so on that have decelerated expansion of development in the said regions.

Importance of Coastal Zone

Bangladesh is an agriculture-based nation and the livelihood of the people of the coastal region is also dependent on agricultural crops, mainly rice. A number of causes affect the usage of land for agriculture in the coastal regions, namely, flooding, drainage congestion, physiography, soil salinity and lack of assured irrigation facilities. Soil salinity is distinctly high during the summer or dry season, thus the harvesting season is restricted to the monsoon or rainy season. Aquaculture is one of the most dominating sources of income for the people of the coastal zone. The *Exclusive Economic Zone* (EEZ) in the Bay of Bengal also supports marine fisheries, mainly shrimp farming. It has been predicted that national development will soon be influenced mainly by the coastal fisheries resources.

Covering about 6,017 square kilometers and situated in the southwest corner of Bangladesh, the Sundarbans is the largest mangrove forest in the world. The Sundarbans is a blessing for Bangladesh and has huge multidimensional use for the local people and also for the nation. The Sundarbans provides sustainable livelihoods for millions of people in the vicinity of the site and acts as a shelter belt to protect the people from storm, cyclone, tidal surges, sea water seepage and intrusion. The area provides livelihood in certain seasons for a large number of people living in small villages surrounding the property, working variously as woodcutter, fisherman, honey gatherer, leaves and grass gatherer among others. The trees in the forest have been very slowly diminishing and one of the primary reasons is deforestation for expansion of agricultural land. Population growth and economic pressure are two prominent factors that have led to the large-scale clearing of forests. In order to remedy the situation, the Forest Department in 1964 took active steps to bring about afforestation in coastal areas. Initially the plantation was done to increase the mangrove shelterbelt which later stretched to government land under the social forestry program.

Industrial areas: The metropolitans, that is, Chittagong and Khulna, and many other cities are situated alongside the coast. Apart from being the second largest city in the country, Chittagong is the main port city and also a commercial capital. The industrial belts in both Khulna and

Chittagong along with ship-breaking yards on the coast of Chittagong are Export Processing Zones (EPZ). According to a report in an English daily, Chittagong generates 40% of the country's industrial output, 80% of its international trade, and 50% of its governmental revenue. Approximately 30 million people in Bangladesh are directly dependent on oceanic fisheries and sea-based commercial transport for their livelihood (Kabir 2019).

Tourism: Most of Bangladesh's present tourist attractions are located in coastal areas. Bangladesh is working towards a sustainable marine tourism industry that will contribute to coastal community development and poverty reduction as well as help protect marine ecosystem and biodiversity. The country offers the rare beauty of sunrise and sunset from the same location in fascinating Kuakata; the beauty of the Sundarbans, the home of the Royal Bengal Tiger; world's longest sea beach in Cox's Bazar; and many more attractions. The coastal tourism is contributing to the economy and generates revenue for the people of the coastal regions. The government is taking measures for the development of tourism sector and its efficient management (see UNESCO 2019; Amin et al. 2019).

Environment: The coastal zone possesses several ecosystems that have important conservation value. These ecosystems are biodiversity hot spots, but planned management is yet to be initiated. Recognizing the gradual depletion of ecosystems and the ecological importance of the flora and fauna, Bangladesh has identified protected areas in the form of national parks, game reserves, wildlife and fish sanctuaries, world heritage sites, marine reserves and ecologically critical areas. There are three wildlife sanctuaries in the Sundarbans and all of them were established in 1977 under the Bangladesh Wildlife (Preservation) (Amendment) Act, 1974, having first been gazetted as forest reserves in 1878. Along with the Forest Act, 1927, the Bangladesh Wildlife (Preservation) (Amendment) Act 1974, controls activities such as entry, movement, fishing, hunting and extraction of forest produce (see Government of Bangladesh 1994, 1999a, 2001b).

Problems of Coastal Zone

Bangladesh is often defenseless in the face of climate change. Its low altitude, high population density and insufficient groundwork work adversely for the nation. The coastal region of the country specifically has the dubious distinction for facing maximum vulnerabilities and natural disasters. Among them, land accretion and erosion in the coastal region generate socio-economic abnormalities and concerns. While erosion forces dislocation of individuals, land seizure and unlawful settlement are predominant in areas where the coast is accreting. Bangladesh frequently experiences cyclones and storm surge which lead to extensive destruction to life and property. The shallow mainland and uniform coastal land aggravate the cyclonic storm surges. Defense from repeated storm surges is an urgent concern in land management (Government of Bangladesh 2001a; Rajja 2019).

Displaced people: Sea level rise, storm, cyclone, drought, erosion, landslide, flooding and salinization are already displacing large numbers of people. It has been estimated that by 2050, one in every seven people in Bangladesh will be displaced by climate change (Environment Justice Foundation 2019). Up to 18 million people may have to move because of sea level rise alone. Approximately 28% of the population of Bangladesh lives on the coast where the primary driver of displacement is tidal flooding caused by sea level rise. By 2050, with a projected 50 cm rise in sea level, Bangladesh may lose approximately 11% of its land, affecting an estimated 15 million people living in its low-lying coastal region (Sarwar 2019).

Due to sea level rise, the progress of salinization has worsened. The backbone of Bangladesh's economy is agriculture, which is being severely affected; harvest is spoiled by increasing salinity, resulting in soil degradation. Consequently, many areas have already been victims of huge harvest losses and substantial decrease in crop and land value. Additionally, the coastal regions are also suffering due to frequency and severity of tropical storms, which cause loss of human life, damage to houses, property and infrastructure and disruption of agriculture and other livelihoods. In 2016 there were four cyclones—Roanu, Kyant, Nada and Vardah—in the Bay of Bengal, although generally there is just one.

Impact on Women: We live in a society where women and children are the most vulnerable groups. During disaster this vulnerability increases. Though disaster itself does not discriminate, the socially construed role of women makes them more vulnerable during the said circumstances. The women of Bangladesh are among the first to face the impacts of climate change and their suffering is disproportionate. The life after a disaster is difficult, the threshold of difficulty increases manifold for women refugees due to multiple causes, like their vulnerability, mental attitude, physical attributes and other social issues. Women who migrate are often at risk of trafficking. It has been observed that the number of Bangladeshi women being trafficked to brothels is rising.

Legal Framework

Natural disasters have made coastal land-use management an important aspect in national development. Hence, coastal land-use management is one of the key features in national coastal development policy and strategy. In 1999a, the Bangladesh Government adopted the Integrated Coastal Zone Management policy. The policy aims to address the liabilities and prospects of the coastal zones, eco-friendly commercial activities and other sustainable use of natural resources. The three main components are a coastal zone policy; a coastal development strategy; and a priority investment program.

The Government of Bangladesh 2004, was introduced for the following purposes:

- Mitigation of natural disasters, safety and protection
- Environment management—defense and rejuvenation of the environment
- Water resources management
- Enhancing rural livelihoods and sustainable economic opportunities for coastal people
- Productive economic activities, intensive growth of tourism and fisheries sector
- Infrastructure development
- Social development including health and nutrition, education, water and sanitation

The Ministry of Water Resources formulated the Coastal Zone Policy, published in the official Gazette in March 2005. The main objectives of the Government of Bangladesh 2005 were:

- Reduction of coastal vulnerabilities and improvement of survival mechanism
- Improving the livelihood of the seaside community by ensuring their basic needs
- Guaranteeing the optimal usage and sustainable management of coastal resources
- Creating an empowering institutional environment
- Advancement of gender equality through women's progression
- Conservation and enrichment of endangered ecosystems

The Government of Bangladesh (2006) was made to organize synchronized priority actions and preparations for their execution via choosing planned priorities and setting targets. The main priorities of Government of Bangladesh (2006) were:

- Ensuring clean and safe water availability
- Protection from human induced and natural hazards
- Improving use of coastal lands
- Stimulating economic growth
- Sustainable management of natural resources
- Betterment of costal livelihood; especially women
- Environmental conservation
- Empowerment via knowledge management
- Assisting institutional environment

Other legislation and policies associated to coastal zone management of Bangladesh primarily comprise of policies and legislation for the management of coastal forests and trees; policies and legislation for the management of coastal land; policies and legislation for the management of coastal water; and policies and legislation for the management of coastal fisheries.

The Government of Bangladesh 1994, highlights the formation of plantations on all recently accreted lands in the coastal zones. It is dedicated to preserving the resources and ecosystems of the Sundarbans. Safeguarding the nation against natural disasters is the primary objective of the national Government of Bangladesh 1992. It concentrates on the conservation of ecological equilibrium and sustainable development. On the subject of coastal forest, the Government of Bangladesh 2005, has a comparable assurance. It usually underlines sustainable development in the coastal zone and hence, straightaway supports the formation of coastal plantations and preservation of prevailing coastal forests and conservation of coastal habitations.

As per the Government of Bangladesh 2001b, generally the coastal "Char Land" is at first given to the Forestry Department for 20 years. During the period, mangrove forests are planted as part of the coastal defense system. As accretion progresses, the land behind the mangrove belt is converted into agricultural land for settlement. Thus, the government also has a limited scale land settlement program. People who have lost their land due to erosion are given the priority in the allocation of such land. The Government of Bangladesh 1992, stimulates usage of land depending on the ecosystem predominant in various regions of the country. It highlights that efforts ought to be reinforced for land recovery, erosion defense, soil fertility and decrease of soil salinity and/or alkalinity. The said policy also indicates reassigning newly accreted land to the Forestry Department on a priority basis to stabilize and defend the land from erosion.

The Government of Bangladesh 1994, declares that "effective measures will be taken for afforestation in the newly accreted char in the coastal areas, as it protects soil and reduces the velocity and intensity of cyclones, tornados and tidal bores". The aforesaid further assists to stabilize newly accreted land. The Government of Bangladesh 1999b, proposes undertaking surveys and inquiries on the problem of riverbank erosion; developing and executing master plans for river training and erosion control; and planning and executing systems for retrieval of land from the sea and rivers.

The Government of Bangladesh 1998b, suggests that every single village should have a pond specifically set aside only for drinking water. The said pond is also to be secured from contamination and surface runoff. The waterlogging problem has produced extensive dissatisfaction amongst the people. The Government of Bangladesh 2001a, reflects upon the importance to rationalize coastal embankment systems on the basis of environmental audit. Last but not the least, the policies and legislation for the management of coastal fishers are the Government of Bangladesh 1983, and Government of Bangladesh 1998a. They are two of the main policies on marine fisheries for Bangladesh.

Conclusion & Recommendations

All these policies have clear implications for coastal development, but in most cases do not have specific sections on coastal areas and often fail to capture the distinctive combinations of vulnerabilities and opportunities that characterize the coast. The policies have been developed for all concerned institutions to put their efforts together for the development of the coastal zone of Bangladesh. Though systematizing their actions in the coastal zone and expanding the foundation for a firm synchronization mechanism. Nevertheless, a lot of improvement is still required for the sustainable development of the coastal zone. The policies were more or less adopted with an ultimate aim to generate circumstances in which poverty would be reduced and the livelihoods of the coastal communities would get developed. Unfortunately, the coastal zone has suffered from various practical management-related difficulties that have led to lower-than-expected economic revenues and more-than-predicted environmental damage.

Therefore, in order to ensure more effective implementation of ultimate policy goals, the policies need to be reviewed and updated. An efficient system for monitoring and impact assessment of policy implementation needs to be developed. Ensuring stakeholder participation in policy formulation is of vital importance. Good management practices need to be followed at all times in order to ensure adequate policy implementation. The policies must include specifically detailed contingency plans for coping with disasters along with guidelines on how such a plan is to be effectively made. The mitigation plans of the problems are also required to be included in the said policies.

References

Environmental Justice Foundation. *Climate Displacement in Bangladesh*. Retrieved on November 30, 2019, from https://ejfoundation.org/reports/climate-displacement-in-bangladesh

Government of Bangladesh. (1983). *Fisheries Marine Fisheries Ordinance*. Dhaka: Ministry of Fisheries and Livestock.

Government of Bangladesh. (1992). *Environment Policy*. Dhaka: Ministry of Environment and Forestry.

Government of Bangladesh. (1994). *Forest Policy*. Dhaka: Ministry of Environment and Forestry.

Government of Bangladesh. (1998a). *National Fish Policy*. Dhaka: Ministry of Fisheries and Livestock.

Government of Bangladesh. (1998b). *National Policy for Safe Water Supply and Sanitation*. Dhaka: Local Government Division.

Government of Bangladesh. (1999a). *Integrated Coastal Zone Management*. Dhaka: Ministry of Water Resources.

Government of Bangladesh. (1999b). *National Water Policy*. Dhaka: Local Government Division.

Government of Bangladesh. (2001a). *Development Strategy of the National Water Management Policy*. Dhaka: Ministry of Water Resources.

Government of Bangladesh. (2001b). *Land Use Policy*. Dhaka: Ministry of Land.

Government of Bangladesh. (2004). *Priority Investment Program*. Dhaka: General Economic Division, Planning Commission of Bangladesh.

Government of Bangladesh. (2005). *Coastal Zone Policy*. Dhaka: Ministry of Water Resources.

Government of Bangladesh. (2006). *Coastal Development Strategy*. Dhaka: Ministry of Water Resources.

Kabir, E. *What's So Special About Chittagong?* Retrieved on November 3, 2019, from https://www.dhakatribune.com/opinion/op-ed/2016/12/23/whats-special-chittagong

Rajja, M. *Coastal Zone Policy*. Retrieved on November 29, 2019, from https://www.thedailystar.net/news-detail-82707

Sarwar, Md. G. M. *Impacts of Sea Level Rise on the Coastal Zone of Bangladesh*. Retrieved on November 10, 2019, from https://www.lumes.lu.se/sites/lumes.lu.se/files/golam_sarwar.pdf

UNESCO. *The Sundarbans*. Retrieved on December 8, 2019, from https://whc.unesco.org/en/list/798

N. Nabila Hoque is a Barrister of The Honourable Society of Lincoln's Inn. She completed her Honours and Masters in Law from University of Northumbria, UK, and is an Accredited Civil Commercial Mediator of ADR-ODR International. She is an Associate Barrister of the Chambers Legal Counsel and part-time Lecturer at British School of Law, Dhaka, Bangladesh. The young researcher works in the field of environmental law, labour law, dispute resolution, public policy and administration.

Disaster Mitigation & Planning for Tsunami in Coastal Areas

Mehul Padharia

Introduction

Gujarat, a coastal state of India, is situated in the shadow of the Western Ghats. Kutch is the most vulnerable district of this state, resembling a tortoise-shaped shallow landmass projecting into the Arabian Sea. Potential hazards, both natural and man-made, distinctively affect the district. The Tropic of Cancer passes through this district, making it more vulnerable to tropical cyclones. The Kutch is a peninsular district of Gujarat state with approximately 750 km of coastline, which makes it prone to flood as well as tsunami. The Makran trench, located off the coast of Iran, is prone to earthquakes and has a potential to generate high-intensity tsunami, which in turn can affect the life and livelihood of Kutch district. Tsunami is unpredictable, hence it is not possible to identify the time of occurrence, location of the epicenter and lead time, prior to the strike of the event, making it very difficult to implement the response mechanism and evacuation process. It is necessary to understand the importance of disseminating early warning signals and spreading knowledge regarding the hazard

M. Padharia (✉)
Gujarat State Disaster Management Authority, Gandhinagar, India

A. Singh et al. (eds.), *Development in Coastal Zones and Disaster Management*, Disaster Research and Management Series on the Global South, https://doi.org/10.1007/978-981-15-4294-7_4

and evacuation process to safe areas. The common understanding is very necessary for administrators as well as the community in order to ensure preparedness against the catastrophic effects of tsunami.

Two most challenging tasks for disaster management authorities are (1) Planning of urban areas and (2) Planning of emergency evacuation to cope with the impending disaster (UNESCO, January 2008).

Planning of Urban Areas

Urban planning has a vital role in disaster preparedness. It includes zoning, building bylaws, identifying zoning restrictions, identification of higher and safer grounds for evacuation, renovation or retrofitting of chronically weak structures. This total exercise is part of advance planning and preparedness. There is need for networking with many other urban bodies and departments. Slack and slow state disaster management authorities (SDMA) may never be able to handle tsunami if they do not work beforehand.

Planning of Emergency Evacuation 'Evacuation Planning' is part of preparedness exercises and is combined with preliminary mapping undertaken by the SDMA. This planning for evacuation becomes disorderly if SDMAs fail to map vulnerable structures, open safe higher platforms and shelters for the vulnerable populations and set priorities for evacuation, which is important in case of unpredictable natural disasters. Some natural hazards such as tsunami give 'lead time' in which proper planning and evacuation mechanism, if implemented wisely, would save many lives.

According to UNESCO, there are five steps towards tsunami preparedness (UNESCO, January 2008) (Fig. 1).

The much-needed early warning technology has been updated and installed in many Indian coastal areas to mitigate the effects of tsunami, but some coastal areas are still highly vulnerable to earthquake-generated tsunami. If a high-magnitude earthquake strikes near coastal areas in the ocean, it would generate a tsunami with very less lead time, which would affect evacuation mechanism thus resulting in a colossal and unprecedented loss of lives.

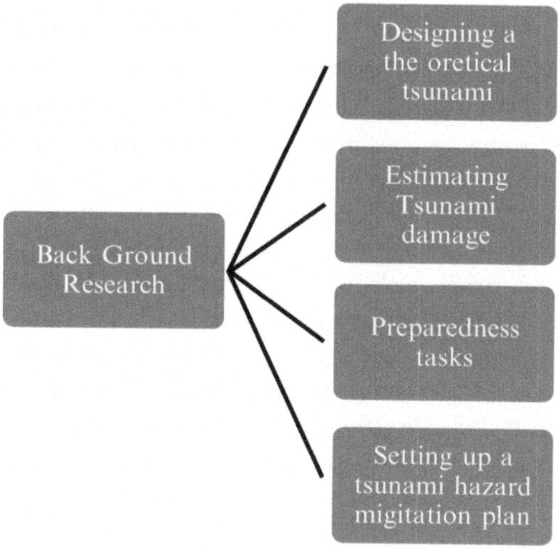

Fig. 1 Five steps towards Tsunami preparedness. (Source: Tsunami preparedness: information guide for disaster planners, 2008)

From Hazard to Disaster: Human Indulgence

Sometimes, an avoidable tsunami becomes a disaster due to human intervention. Destruction of natural protection such as levees, coastal trees, mangroves and ecological areas due to construction of tourists resorts and hotels in the onshore areas contribute to increasing the intensity of the tsunami strike leading to the vital loss of life (TERRA.WIRE 2004). In Europe, many trees were chopped off along with scraping of mangroves for good shrimp production. But since coral reefs and mangroves help to dissipate the strength of killer tidal waves, their absence caused much damage in the region (EJF 2006). Destruction of natural wetlands, deforestation and urban development detain the flow of flood water from inland to coastal area after Tsunami strikes. The water inundates the area as it unable to escape (GradeMiners 2017). If this inundated water remains in inland for long, it can contaminate drinking water. In addition, it can also lead to the spread of waterborne diseases such as malaria and cholera rise due to presence of still water.

Survival in Tsunami 2004

More than 2,00,000 lives have been lost in the Indian Ocean Tsunami of 2004 (Express web desk 2019). The occurrence of a massive earthquake with an epicenter off the coast of Indonesian Island of Sumatra inundated the area within a few minutes. Information dissemination and evacuation process were delayed as the threat information was received only a few minutes before the lead time. In a short span of time, one cannot reach a safe place without understating the characteristic of the impending hazard. Orientation and sensitization programs would have helped to increase the coping capacity of the local communities through knowledge building. Understanding the properties of hazards may help in faster evacuation, for example, if a high magnitude of the earthquake is being felt in the coastal area, the local people must understand the potential of the tsunami threat which would strike the coast in a few minutes. This knowledge will lead people towards a safe area without any official warning and would save substantial damages and losses. Integrated public education and planning for preparedness is needed to make this knowledge generic (Orcutt 2010).

The high intensity of tsunamis such as Indian Ocean Tsunami of 2004 has the potential to strike the areas which were far away from the epicenter. The waves of this type of tsunami surpassed the epicenter and reached far flung areas of coastal region to spread havoc, but a good lead time would have helped to disseminate the official warning message to the low coast area. An adequate knowledge of the hazard can help in devising an early warning system and wielding golden time period for evacuation.

Improving Disaster Measures in Gujarat

The state of Gujarat is situated in the western part of India, with the longest coastline of 1600 km (John 2018), which makes it vulnerable to coastal hazards. There are 41 ports in which 1 major, 21 intermediate and 29 minor ports which reduce resilience of this coastline in many ways. Gujarat State Disaster Management Authority (GSDMA) has taken significant steps to increase the coping capacity of coastal communities. Various types of mock drills and evacuation exercises are being conducted periodically to spread the awareness among local communities. This type of regular exercise helps in building disaster resilience among communities and is also helpful in finding the gaps and drawbacks in Incident Response

System of administration towards incremental reforms (gradual reforms in small packages from time to time learning through various disasters).

IOWave18 mock exercise or the Indian ocean-wide tsunami **mock exercise** was conducted recently in Gujarat with the support of Intergovernmental Oceanographic Commission (IOC) of UNESCO, NDMA, ITEWC and INCOIS to find out the flaws in warning dissemination system. It includes 23 other nations who have participated to learn evacuation of thousands of people from coastal areas and over half a dozen states within India. This exercise was conducted in two phases: the first phase was conducted on 4 September, 2018, in the western coast region and the second phase was conducted on 5 September, 2018, in the eastern coast region. Two districts from the western coastal region, Kutch and Jamnagar, were identified for the exercise. Kutch which is a peninsular district of Gujarat with 750 kilometres of coastline which makes it prone to a higher intensity of coastal hazards. Mandvi taluka of Kutch district was identified as the staging area. A hierarchy of Incident Response System (IRS) was recognized and instructions were given by the District Collector to all concerned officials to strictly follow the IRS system. Various Integrated Task Forces (ITF) were created for search, rescue and evacuation process. Staging area was identified in high MSL area which would be safe to reach during a catastrophic event. Various types of posts such as a 'Responsible Officer Post' followed by many others such as the Incident Command Post, Operation Section Post, Planning Section Post, Media Management Post, Lost & Found Post, Relief Camp for the victims, Medical Post for the injured, Water Supply Post, Food Unit Post, Generator Set post were set up in the Staging Area.

Integrated Task Force consists of the following: a Task Force Commander who leads the Task Force to the casualty site; home guards and police officials to maintain law and order; civil defense and volunteers of non-governmental organizations for evacuation; national and state disaster response forces (NDRF and SDRF) which are equipped with high-tech machinery that would help in proper search and rescue operations; electrical department which would help in the removal of damaged or fallen electric polls and Nagarpalika (local authority) officials with excavation machinery which would help in the removal of debris from the route of 'Task Force'. These task forces consist of highly qualified officials, which would help in spreading awareness among the local public during the mock exercise (Fig. 2).

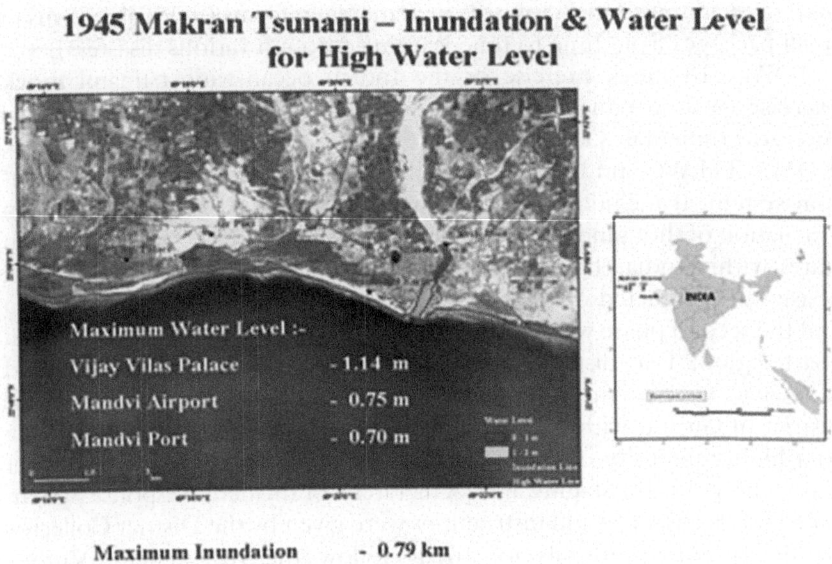

Fig. 2 Inundation area of Mandvi Taluka from Kutch district due to 1945 Makran trench earthquake and tsunami. (Source: Ministry of Earth Sciences, March 2009)

Integrating Disaster Education

Many countries are taking significant steps in the field of education for Disaster Risk Reduction (Petal 2008). The youth are the key to spread awareness among the communities and therefore, inclusion of disaster education program in school curricula may sensitize students to understand the meaning of early warning systems and how to respond during a disaster. School Safety Week Program is a good initiative taken by the National Disaster Management Authority (NDMA) to celebrate disaster risk reduction program in all schools (National School Safety Programme (NSSP)—A Demonstrative Project n.d.). Students are informed about disasters by following concepts of Learn with the Fun with the help of sensitization and orientation programmes through mock drills, games and rallies. The Disaster education program is an effective exercise when it is conducted in a community, involving people from all walks of life including female groups, pregnant women, handicaps, older and weak people, children, shepherds or owners of livestock and the marginalized. The evacuation process is different for each group, such as for children and women, aged, weak and the destitute. There would be some energetic participants who would be able to climb more comfortably than others and reach the identified spot during airlift operation while the weak, disabled, elderly and pregnant women would be relatively slower or may not be able to reach the evacuation spot. Therefore, it is necessary to identify a specific spot or ground where airlift operations can take place to save lives. Sometimes shepherds and livestock owners and subsistence farmers lose their lives as they are not able to evacuate the place on time with their basic resources. For them, a specific and different route could be identified away from jammed roads for a proper evacuation during the mayhem of rapid evacuation.

Tourists are clueless during disasters because of the absence of their guides and also language barriers. They are not well versed with the topography and demography of the local area. Pamphlets of evacuation routes must be distributed in the tourist areas which are easily accessible and landmarks for gathering must be identified. Proper signages must be provided at prominent locations. Hotel owners and tourist guides must train their staff to assist tourists in the evacuation process during a response phase.

Tsunami Protection Barriers and Measures

In a densely populated coastal area, structural mitigation measures have a powerful barricading role along with the non-structural mitigation measures. To defend the coastal life from a killer tidal wave, strong man-made as well as natural protection lines are needed. These defence lines can extend from hinterland (building structures) to the onshore (reefs and levees). There are basically two types of natural tsunami barriers: offshore and onshore. Sand and coral reefs are types of offshore barriers. The research has spelled out the importance of these barriers as the coastal areas, which consist of dense coral reefs, have experienced very less or no devastation (UNEP 2005). Mangroves, sand dunes and beach tree belts and forests are the onshore barriers. The resilient and dense foliage vegetation helps to protect the coastal area from tsunami waves according to the UNEP-report (2005) and IUCN-reports (2005).

The construction of evacuation buildings in densely populated coastal areas is of paramount importance, especially where suitable multi-storied buildings are not available for airlift operations. A proper design of evacuation is necessary which may be beneficial against devastating tsunami waves as well as during earthquakes with a magnitude of 10 (Oumeraci 2006). Four-storied evacuation building constructed in Banda Aceh are good measures for airlift operations. The height of this building is more than 18 m with no partitions or hollow structure, following the concept of mosque construction design. This technique may sustain the wave force during tsunami and heavy columns can withstand earthquakes of magnitude 10. (Suppasri et al. 2015).

Conclusion

This chapter contains experiences and best practices from administrators. With proper planning and effective preparedness, the consequences of tsunami may be mitigated. It is through Hazard Risk Vulnerability assessment, coordination between the governmental and non-governmental organizations, disaster education sensitization programme, early warning, forecasting and proper information dissemination that the losses due to tsunami can be significantly reduced. It is seen that the collective efforts of the authorities and citizens are required for effective disaster preparedness and mitigation. A sustainable endeavour towards resilient communities and experienced administration is necessary to stay prepared against any impending disaster which may strike any time and destroy many generations of progress and togetherness.

References

Coast as a Part of the Early Tsunami and Storm Surge Warning System, National School Safety Programme (NSSP)—A Demonstrative Project. (n.d.). Retrieved from National Disaster Management Authority https://ndma.gov.in/en/national-school-safety-programme-a-demonstrative-project

EJF. (2006). *Mangroves: Nature's Defence Against Tsunamis—A Report on the Impact of Mangrove Loss and Shrimp Farm Development on Coastal Defences.* London: Environmental Justice Foundation.

Express web desk. (2019, November 20). 'December 26, 2004 Tsunami: 12 Years on', New Delhi, 1:18:00pm. Available at https://indianexpress.com/article/india/december-26-2004-tsunami-india-thailand-indonesia-srilanka-natural-disaster-earthquake-12-years-on-4445361/. Retrieved 1 Feb 2020.

IUCN. (2005). Mangrove Forests Saved Lives in 2004 Tsunami Disaster. 19 December 2005. http://www.iucn.org/tsunami/

John, P. (2018, January 28). Sea Grabs Lands of Gujarat Eroding. *The Times of India.* 05:51 IST. Available at https://timesofindia.indiatimes.com/city/

ahmedabad/sea-grabs-land-gujarat-coast-eroding/articleshow/62677706.cms. Retrieved 20 Jan 2020.

John A. Orcutt, M. R. (2010). *Tsunami Warning and Preparedness, National Research Council.* Washington, DC: The National Academies Press.

Ministry of Earth Sciences. (2009). Modeling and Mapping of Tsunami Along Indian. Delhi: Government of India.

Nita, J. (2017, July 14). How Human Activities Can Have an Impact on Natural Disasters. GradeMiners. Blog available at https://grademiners.com/blog/three-ways-in-which-human-activities-can-have-an-impact-on-natural-disasters. Retrieved 2 Feb 2020.

Oumeraci, H. (2006). *Near- and Onshore Tsunami Effects. Knowledge Base Generation and Model Development.* Background Paper for DFG- Round Table Discussion 2007, Hannover, Germany, 19 pp. Available at http://www.fzk.uni-hannover.de/323.html

Petal, M. Ch. 6.1 Disaster Risk Reduction Education. In R. Shaw & R. Krishnamurty (Eds.), *Disaster Management: Global Challenges and Local Solutions.* Universities Press, India. 2008 *(18) (PDF) Education in Disaster Risk Reduction.* Available from https://www.researchgate.net/publication/312489763_Education_in_disaster_risk_reduction. Accessed 19 May 2020.

Suppasri, A., Goto, K., Muhari, A., Ranasinghe, P., Riyaz, M., Affan, M., Mas, E., Yasuda, M., & Imamura, F. (2015). A Decade After the 2004 Indian Ocean Tsunami: The Progress in Disaster Preparedness and Future Challenges in Indonesia, Sri Lanka, Thailand and the Maldives. *Pure Applied Geophys, 172,* 3313–3341, Open access at SpringerLink.com. https://doi.org/10.1007/s00024-015-1134-6.

TERRA.WIRE. (2004, December 27). *Human Activities Contributed to Tsunami's Ravages: Environmental Expert.* Paris. Retrieved from Terra Daily http://www.terradaily.com/2004/041227155435.4ap75nje.html

United Nation Environment Programme (UNEP). (2005). *After the Tsunami—Rapid Environmental Assessment* (pp. 301–304). Nairobi: UNEP.

UNESCO report prepared by (2008). *Tsunami Preparedness: Information Guide for Disaster Planners.* Intergovernmental Oceanographic Commission [5321] Document code:SC.2008/WS/5 (Rev. only in Fre). Paris: UNESCO.

Mehul Padharia is a District Project Officer of Kutch District working for Gujarat State Disaster Management Authority. He is a trained architect from MANIT Bhopal and a post-graduate in disaster mitigation and management from Indian Institute of Technology, Roorkee. He has been actively working in the government sector for disaster management and for building a disaster-resilient society. He has been involved in planning and conducting trainings workshops on disaster risk reduction and mock-drills such as Mega State Level Mock Exercise & IOWave18 International level Tsunami Mock Exercise for capacity building and public awareness generation, for which he has also received awards.

PART II

Land, Agriculture and Food at the Coastal Rim

Coastal Agriculture and Future Challenges

Anurudh K. Singh

Introduction

Coastal areas are commonly defined as the interface or transition areas between land and sea, including large inland lakes. Favorable biophysical and climatic conditions, together with ease of communication and navigation, have encouraged human settlement in coastal areas since prehistoric times. They are also important ecologically, as they provide numerous environmental goods and services. The coastal areas contain both critical terrestrial and aquatic habitats, particularly in the tropics. Such habitats together comprise unique coastal ecosystems, supporting a rich biological diversity and valuable natural resources. Examples of such habitats are estuarine areas, coral reefs, coastal mangrove forests and other wetlands, tidal flats and seagrass beds, which also provide essential nursery and feeding areas for many types of coastal and oceanic aquatic life. It is estimated that 90% of the world's fish production is dependent on coastal areas. In

A. K. Singh (✉)
Division of Germplam Conservation, National Bureau of Plant Genetic Resources, Indian Council of Agriculture Research, New Delhi, India

International Crops Research Institute for the Semi-Arid Tropics (ICRISAT), Patancheru, Telangana, India

A. Singh et al. (eds.), *Development in Coastal Zones and Disaster Management*, Disaster Research and Management Series on the Global South, https://doi.org/10.1007/978-981-15-4294-7_5

addition, coastal areas support large numbers of migratory and non-migratory waterfowl and shorebirds, and endangered reptiles, such as turtles and alligators. The economic advantages of using and maintaining coastal biological diversity have been recognized since the inception of human settlement and culture.

Importance of Coastal Agriculture

The coastal areas often provide excellent soil and climatic conditions for agriculture, which has been practiced for thousands of years and plays an important role in the economy of coastal areas. Aside from providing food to coastal populations, agriculture also often provides raw materials to industry, which is established in these areas to make the best use of port facilities. Therefore, coastal agriculture, in addition to benefiting from favorable environmental conditions with generally good land, also benefits from sea communications for trade and for development of industry and tourism in coastal areas, which may provide markets for agricultural products, livelihood support for coastal populations, including cities, and opportunities for establishment of agri-based sectors.

In some coastal areas and islands, agricultural production makes an extremely important contribution to the local economy or to the national agricultural production. In countries such as Egypt, India and Bangladesh, the river deltas, with their fertile alluvial soils, play a major role in the agricultural sector (Bacow and Wheeler 1984). India has one of the world's longest (approximately 7500 km) coastlines, characterized by varied landforms and ecosystems (Fig. 1 depicts the coastal areas of India). In India, agriculture and allied sectors share 15.87% of total Gross Value Added (GVA) and employ 59% of the total workforce (Ministry of Statistics and Programme Implementation, 2020), to which 0.91% and 5.23% GVA comes from fisheries and coastal agriculture and allied sectors respectively. Even where agriculture may not appear to be significant in terms of contribution to the local economy or to national agricultural production, there are several reasons for providing specific attention to coastal agriculture to make it more productive and attractive and for integration in the coastal area plans to face future challenges of food and nutritional security for an ever-growing human population and climate change. These are as follows:

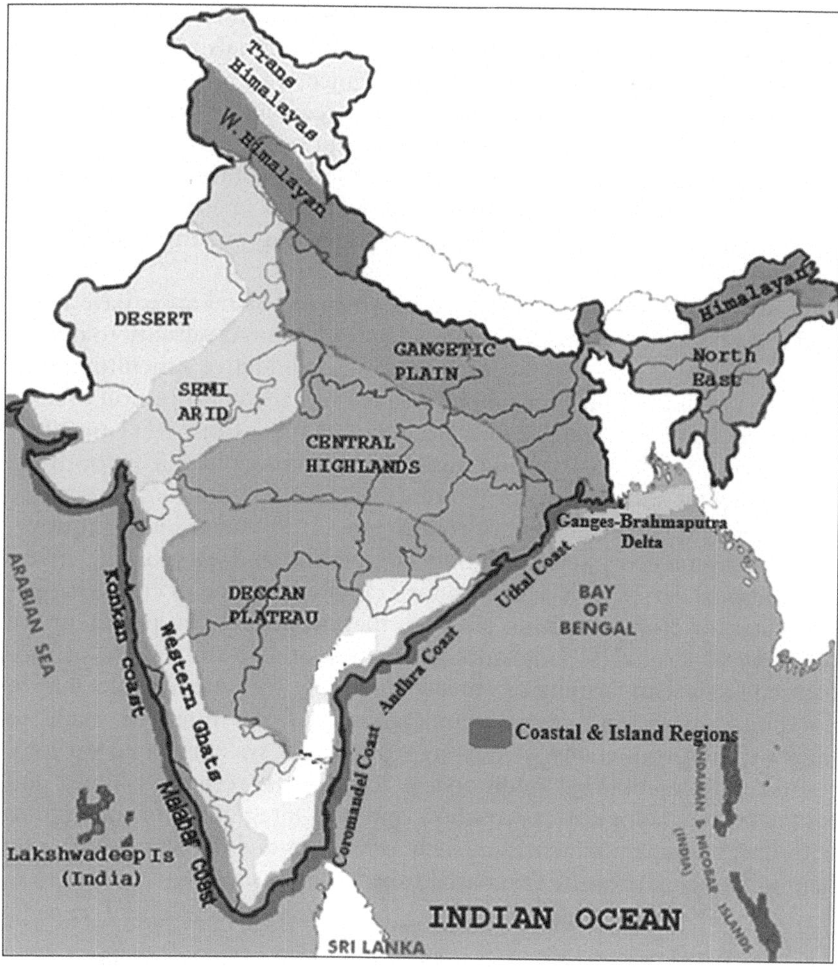

Fig. 1 Coastal agriculture and future challenges. (Source: A restructured version of "Map of India displaying geographic regions as effecting its bio-diversity" available on public domain under slideshare.net)

- Agriculture plays a key role in the local economy, either through the production of food or by providing raw materials to industry, and thus has strategic and political significance.
- In many developing countries a large percentage of the population is dependent on agriculture for their livelihoods.
- Agriculture is the main user of land, and agricultural activities can have a significant impact on natural resources in a coastal area, particularly on the quality and flows of water and on natural habitats.

However, recent capital-intensive activities in coastal areas have made land tenure fragile for smallholder farmers with limited capacity to defend their interests, or they may be tempted to sell productive agricultural land for non-agricultural purposes. For these reasons the contribution of agriculture to GDP has been steadily declining in India from 1951 to 2011. This situation requires attention during the planning phase from both the environmental and food security points of view. In addition, sustainable and eco-friendly management of coastal agriculture can facilitate improvement of coastal ecosystems, avoiding competitive and antagonistic effects of traditional agricultural activities, for example adoption of crops that use less water, or crop protection practices that do not rely on persistent use of chemical fertilizers, fungicides, pesticides, and so on, or use of submerged lands with farming of water-loving crops with greater tolerance to flooding, salinity and higher temperatures. Efforts must be made to improve the productivity of existing agricultural systems based on local resources by combining traditional practices with recently evolved and tested biotechnological approaches, and the integration of agriculture with other companion activities, such as fisheries and forestry, which may help in reducing pressure on coastal fisheries and wetlands.

Some Important Agricultural Systems of Indian Coastal Regions

Many agricultural systems around the world have been developed in coastal areas as per the opportunities offered by the prevailing resources, arising from the location and demand for food or raw agricultural materials for industrial growth. For example, coastal areas consist of alluvial accumulation plains with generally deep, relatively flat, fertile soils and benefit from a substantial supply of water, from surface and/or subsurface sources, offering greater agricultural output, if appropriately exploited. A milder and more humid climate, prevailing in coastal areas, may favor growth of a crop or crops not grown elsewhere in the country. Diversity of coastal environments, which include both terrestrial and marine ecosystems, may facilitate a combination of agriculture, pisciculture and forestry from which coastal populations draw their livelihoods, and sometimes also from seasonal work in the recently developed tourism sector, globally. We discuss here some coastal agricultural systems that have evolved over time in India, which are predominantly agricultural-based, through the ingenuity of local communities facilitating effective and beneficial agriculture, and integrating diverse crop farming, forestry, livestock, poultry and fisheries, and effective management of prevailing natural resources.

Traditional Rice-Based Farming System of Lower Gangetic Plains/Delta This involves rice (*Oryza sativa*) cultivation both in rainy (*kharif*) and post-rainy (*rabi*) seasons, generating diverse production systems suited to diverse and difficult ecologies. This has led to the evolution of a wide genetic diversity of traits desirable for genetic engineering of existing cultivars with improved resilience and productivity. More than 5500 varieties of rice are reported from Bengal (Deb 2000). As per ecological adaptation, rice can be grown on marginal land ranging from deep water to dry scanty rainfall areas (Singh 2012, 2015). Under traditional cultivation, rice is cultivated on high, middle and lowlands, and farmers have evolved varieties suited to these conditions. The specific agroecology of excessive water during the rainy season and ingress of salinity from the sea has generated a great deal of genetic diversity in rice for traits such as submergence tolerance, salinity tolerance and deep-water floating types (Siddiq et al. 2006). Similarly, for winter cultivation in rainfed swampy areas or low-lying or medium lands with irrigation, a series of varieties of *Boro* rice have been developed. *Boro* is a Bengali word derived from the Sanskrit *Borob*, which refers to a special type of rice for cultivation on residual or stored water in low-lying areas after the harvest of *kharif* rice.

Jute-based Cultivation System of Lower Gangetic Plains/Delta Jute (*Corchorus capsularis*; *C. olitorius*) is a water-loving plant species, and the humid climate of the Ganga-Brahmaputra Delta Region favors its growth. The soil of the Gangetic Delta creates an effective platform for growing jute and other allied fiber crops, such as sun hemp, as rich loam and riverine silts make the best composition of soil and mineral structure for growing jute. Each year, additional silts are brought in by floods, a devastating phenomenon that has been found highly favorable to producing jute and other fiber crops. Three areas of Greater Bengal, (i) *Brahmaputra Alluvium*, also called the *Jat* Area, which is inundated and replenished every year by fresh alluvial deposits, with acidic soils, but produces the best quality jute mostly in the Bangladesh region; (ii) *Ganga Alluvium*, also called the *District* Area, which has alkaline soils, and produces a quality of jute next only to *Jat* jute in order of quality; and (iii) *Teesta Silt*, also called the *Northern* Area, which has sandy soil with lower moisture retention capacity and therefore produces an inferior quality of jute.

Integrated Triple-Cropping System of Lower Gangetic Plains/Delta The availability of moisture for the greater part of the year has permitted the evolution of a triple-cropping system. This system includes the rainy (*kharif*) season from June to October; a winter (*rabi*) season from November to February; and a summer (*dry*) season from March to May. It may involve rice during kharif, pulse crops (food legumes) in winter, and vegetables in summers. The major cropping systems practiced are rice–wheat, rice–potato, jute–wheat, jute–rice–sunflower, jute–mustard, rice–mustard, jute–rapeseed–rice, sesame–groundnut– sunflower, pulses–wheat, rice–vegetables, and rice–wheat–summer rice, evolved by local farmers based on their ingenuity. These cropping systems are being further evolved. Biswas et al. (2006) concluded that the new triple-cropping system, involving potato, and the traditional systems, such as jute–wheat, jute–rapeseed–rice, and rice–wheat, require fewer inputs and involve less risk. In crop rotation, rice or wheat is taken after rice/jute. The availability of moisture throughout the year has made horticulture and floriculture an important component of coastal agriculture. Several of these systems are common in the Eastern Ghats/coastal regions of India as well.

Lowland Rice Cultivation Under Improved Management of Cauvery River Delta Cauvery River flows from the northwest to the southeast and drains vast areas of the southern Indian Peninsula. The Cauvery Delta and the riverine beds in the plains have accumulated loamy river alluvial soil and, as a result, its eastern delta area is the most fertile. Farmers follow irrigated farming along the hinterland of the Cauvery Basin and Delta with paddy as the major crop. It is a granary of paddy, particularly in areas supported by the network of irrigation canals that includes the eastern coastal or deltaic regions of Tamil Nadu, that is, Thanjavur, Nagapattinam, Cuddalore and Pudukottai. The Cauvery River, in its course through Karnataka and Tamil Nadu, is interrupted by 12 man-made dams ('anicuts') diverting water through artificial channels for the purpose of irrigation and drinking. The dams constructed using earth and stone have borne the destructive force of nature for centuries and protected regional life and livelihoods (Raman 2008). Most of the distributing canals of the Cauvery River crisscross the deltaic region. Many tanks were constructed during the Chola period, for example Rajendra Chola I's huge tank, called Solagangam in Gangaikonda Solapuram.

The oldest major dam on the Cauvery River was built in the second century CE at Kallanai, Thanjavur (Grand Anicut Dam) for efficient management and use of water resources to support agriculture, ensuring food and nutritional security for the region. It was one of the first dams to be constructed in the world. The Grand Anicut is believed to have been built in the mid- to late Chola period and was later rebuilt by the British. It was a vast construction for its time, spanning more than 300 m across the Cauvery, with a width of more than 18 m. The Cholas controlled this system by granting water rights, just as they would grant land rights to kinsmen, military chiefs, royal retainers, village officers and especially religious institutions (Raman 2008). The area was defined as the 'Rice bowl of Tamil Nadu', where the fan of water channels in the Cauvery Delta was used to irrigate fields, thereby providing almost constant water supply to the lowlands. The drainage network of the river is dense, and the river forms a delta at Trichinopoly (Tiruchirappalli, Karur, Ariyalur and Perambalur). More than 60,000 ha of the cultivable lands are irrigated by these channels in Mandya, Mysore and Thanjavur districts. Rice is the main crop of the region and is traditionally grown under both rainfed and irrigated agroecosystems with maximum genetic variability, including for traits such as high temperature (heat tolerance), to respond to the major challenge of raising crops under rising temperature due to climate change with global warming (Singh 2013).

The Tamil Nadu Coastal Fishing System The Cauvery River system is very rich in fish biodiversity (Jayaram et al. 1982; Balasundaram et al. 1999).

The coastal areas as a continuation of the river basin have fishing as a major agricultural activity. The coastal fishing communities in Sirkali Taluk have used the *Catamaran* for the last 2000 years. The *Catamaran* is a sustainable eco-friendly traditional fishing system crucial for preserving and promoting sustainable fishing in coastal areas. It is a type of double-hulled boat or ship consisting of two hulls or *vakas*, joined by a structure, the most basic being a frame, formed of *akas*. It has evolved into a light watercraft called *Kattumaram* in the Tamil language. The word *kattu* means 'tie', and *maram* is the 'wood tree', and thus *Kattumaram* simply means 'tying two trees together'. It was invented by the *Paravas*, an aristocratic fishing community of southern Tamil Nadu. Recently, this system has also been considered by the FAO as one of the Globally Important Agriculture Heritage Systems (GIAHS) (FAO 2008).

Ulkatni and Awatni Rice Cultivation in Konkan Region These are two local practices followed in the *Khar* (salty) lands of Thane and Raigad districts in Maharashtra. In *Ulkatni*, the clods are turned upside down with the help of a crowbar in the months of April–May, while in *Awatni* the rice seedlings are planted in the field along with the mud ball from the seedbeds (transplantation) after tillage following the harvest of *kharif* rice in reclaimed *Khar* lands. *Awatni* is significantly superior to the regular practice of transplanting, provided the plant population is maintained. Superficial planting in *Awatni* avoids contact of the tender seedlings with the salty portion of the soil, thus avoiding seedling mortality (Singh 2014a).

Mixed Cropping System of Konkan Region A tropical climate with high rainfall has supported the cultivation of horticultural crops, including both rainfed and irrigated, fruit and plantation crops, and cereals, pulses, vegetables and spices. For example, papaya is intercropped in sapota orchards; pineapple in coconut orchards; spices such as black pepper, cinnamon, cardamom and nutmeg with coconut and areca nut; and clove and other spices with coconut. Coconut is a smallholder's plantation crop, cultivated as an allied crop in a varied crop-mix. Approximately 92% of the total area of the region is under coconut farming. Cashew nut is one of the traditional crops of the region, mainly grown on hill slopes in coastal areas as a rainfed perennial horticultural crop.

Fishing System of Konkan Region Farmers in the Konkan region use two methods of fishing, a traditional method called *Dol* and the other using trawlers. In the *Dol* method, hand-woven nets are used for catching the fish. The nets are placed 18 feet deep in the water with the help of wooden poles and two boats, which help adjust distances of the net. The net is pulled off one hour before high tide. The fish, trapped during tides, are then harvested.

Coconut-Based Cropping Systems These systems extend in the Western Ghats from the northern Western Ghats (Konkan) to the southern Western Ghats (Malabar). In these systems, coconut is the major crop intercropped with both annual and perennial crops, such as pepper, areca nut, cocoa, banana, turmeric, ginger, small tubers, and fodder, and in some areas with upland rice, pulses, and oil seeds.

Rice-Based Cropping Systems (In LowLands) of Malabar Region Either a single crop or two crops of rice are grown, depending on the availability of water, as in the central areas of the region, or after dewatering of impounded water, as in the *kayal* lands of Kuttanad region, Alappuzha and Kottayam districts in Kerala. In some areas, vegetables, pulses, and oil seeds are grown in fallows or as a summer crop while in others rice is intercropped with diverse horticulture crops. Fish farming or prawn culture integrated with rice is practiced in the coastal areas of Kerala (Sasidharan et al. 2012). Leena Kumari (2012) listed as many as nine distinct rice-based cropping systems.

'Pokkali' and 'Kaipad' Rice Cultivation Systems in Coastal and Below Sea Level Areas of Malabar Region Under these systems, the surplus water resulting from excessive rains and regular action of sea waves and shore currents in the coastal areas and in the areas below sea level has been harmoniously and profitably managed with successful rice cultivation by local populations. This includes the practice of rice cultivation in waterlogged areas of southern coastal Kerala (Alappuzha), called *Pokkali*. It is a unique cultivation system of salinity-tolerant rice varieties cultivated organically in the waterlogged coastal regions. It has been further improved with integrated rice–shrimp/prawn farming (Singh 2014b). One rice crop

followed by shrimp/prawn capture provides a substantial subsidiary income to farmers (Jayan and Sathyanathan 2010). This system is also practiced in Thrissur and Ernakulam districts. In Kannur district of North Kerala, a similar system called *Kaipad* is practiced with the rich biodiversity of flora and fauna, organically rich soil, mangroves, and migratory birds. It differs from *Pokkali* in the way it is carried out, which is purely natural relying on the monsoon and the sea tides. In addition, it has its own salinity-tolerant landraces and high-yield varieties of rice (Vanaja 2013). *Kaipad* is also practiced in Kasaragod and Kozhikode districts. These systems together have been designated by the FAO as GIAHS, the Kuttanad rice cultivation below sea level (www.fao.org/giahs/giahsaroundtheworld/designated-sites/.../kuttanad...system/en/).

Homestead Farming Systems The traditional system of home gardens in coastal areas, as per the agroclimatic conditions, favors growing of a wide variety of crops. In this system, farmers choose their crop combinations and integration of livestock or fish farming as per the prevailing conditions. Several intercrops are raised, resulting in a multi-story cropping pattern with canopy stratification. Perennial tree crops such as coconut, areca nut, jackfruit, mango, cashew, tamarind, and forest tree species occupy the upper layer; pepper, clove, nutmeg, cinnamon, cocoa, and so on occupy the second layer; banana, cassava, yam, cocoyam, and so on occupy the third layer; and ginger, turmeric, pineapple, vegetables, grain legumes, and guinea grass occupy the ground layer. This is very similar to the tropical rainforest structure of diverse species (Shehana et al. 1992). The Malabar coastal region is the main producer of spices, which together form the number one cash crop. According to Salam et al. (2008), several home garden systems practiced in the Malabar region are based on biological and physical factors.

Commercial/Cash Cropping System In addition to spices, many cash crops, such as coconut, oil palm, rubber, coffee, areca nut or betel nut, cashew nut, and betel vine are grown on an extensive scale in large contiguous areas of coastal regions as plantation crops. To facilitate greater land use, as mentioned above, the base land is often used for cultivation of other herbal crops in mixed cropping systems.

Further, consequent to the cultivation of diverse crops and farming systems, such as rearing of livestock for milk and meat and poultry, many combinations of integrated agricultural systems have evolved in the coastal areas to make the basal system more profitable. For example, Torane (2009) found 18 major farming systems being practiced in different areas of the Konkan region, as per the micro-climate and resources: 1. paddy + irrigated plantations + dairy, 2. paddy + grass + dairy + goat rearing, 3. paddy + irrigated plantations + flowers, 4. irrigated plantations + dairy, 5. paddy + irrigated plantations + betel vines, 6. paddy + other cereals + rain-fed plantations + dairy, 7. paddy + other cereals + dairy + goat rearing, 8. paddy + pulses + dairy, 9. paddy + vegetables + poultry, 10. paddy + grass + dairy, 11. paddy + vegetables + dairy, 12. paddy + dairy, 13. paddy + dairy + goat rearing, 14. paddy + pulses + dairy + poultry, 15. paddy + rainfed plantations + dairy, 16. paddy + irrigated plantations + rainfed plantations + dairy, 17. paddy + irrigated plantations + vegetables + dairy, and 18. paddy + irrigated plantations + rainfed plantations + poultry.

In addition, coastal agriculture is also supported by other natural resources such as forests and aquatic biological diversity. Coastal forests are mangroves, savannah woodlands, dry forests and rainforests. Commercial exploitation of mangroves provides building materials, such as poles and timber, and numerous non-forest products. Savannah woodlands and dry forests provide grazing land for livestock, and rainforests provide commercial timber. Mangroves are primarily of environmental and social value, providing environmental services, unique habitat for hosting biological diversity, and spawning and nursery grounds for many species of commercially valuable fish. Mangroves also act as a sediment trap, a source of nutrients for inshore waters, and protect shoreline erosion and prevent surge-tide damage (Barbier 1990). Fisheries contribute a significant proportion of GDP of coastal regions, as a large proportion of the total fish production of many coastal regions is derived from coastal fisheries (Barbier 1992). Therefore, sustainable development of coastal fisheries, which is considered one of the components of larger agriculture, determines the ability of many coastal communities to survive, as it contributes to livelihood support for the population, and food security. Coastal areas also provide habitat for approximately 90% of marine fish accrued for commercial and subsistence fisheries. For example, fish farming in combination with rice is an important source of livelihood for farmers in the Bengal Delta, formed by three large river systems, the Brahmaputra, Ganga, and Yamuna. The *Pokkali* and *Kaipad* rice cultivation systems of the Malabar region have been further strengthened with integrated rice–shrimp/prawn farming (Singh 2015). However, recent commercial prawn culture has been a bubble and burst story.

Major Agricultural Crops of Coastal Areas

The availability of moisture throughout the year with excellent soil and climatic conditions facilitates cultivation of many plant species as part of coastal farming/agriculture. Some important ones under various crop groups are listed here for ready reference:

Cereals: Rice, wheat (Gangetic Delta)
Grain Legumes: Grass pea, horse gram, lentil, *Vigna* spp.
Forage Crops: Tropical grasses
Fiber Crops: Jute and allied fibers (flax, kenaf, ramie)
Oilseeds: Mustard, rapeseed, sesame, sunflower
Vegetables: Amaranth, *Arvi* or taro (*Colcasia*), basales, brinjal, cassava, drumstick, Indian spinach, okra, potato (Gangetic delta, recent introduction), Vigna beans, water spinach, yam
Fruits and Nuts: Jackfruit, mangosteen *(Garcinia mangostana)* and *Garcinia* spp., papaya, pineapple (Vazhakulam pineapple), sapota, toddy palm (*Borassus flabellifer*)
Spices and Condiments: Black pepper, cardamom, cinnamon, clove, ginger, Malabar tamarind, nutmeg, turmeric, vanilla
Plantation Crops: Areca nut, banana, cashew nut, cocoa, coconut, coffee, oil palm, rubber
Medicinal and Aromatic Plants: *Jatamansi* or muskroot (Gangetic delta), turmeric (*Curcuma* spp.), lemon grass, *Rauvolfia*, *Piper longum*
Ornamental Crops: *Chrysanthemum* spp., crinum lily, doum palm (*Hyphaene thebaica*), ginger lily, ixora (*Ixora coccinea*), jacaranda, jasmine, lotus, palms
Forestry/Agroforestry: Bamboo, Malabar mahogany, Indian rosewood, *kambakam* (*Hopea ponga*), rosewood (*Dalbergia* spp.)
Others: Betel leaf, mangrove date palm, rattans (*Calamus* spp.)

Table 1 lists vegetable crops grown in the coastal areas of Oregon, USA, reflecting the opportunities offered by coastal conditions with high moisture content in a small coastal area, for cultivation of a wide range of both tropical and temperate crops.

Table 1 List of vegetable crops grown in coastal regions of Oregon, USA, for direct sale[a]

Crops	*Status*
Artichokes[b]	Well adapted.
Snap bean	Commercially grown only in small plots
Beets	Commercially grown only in small plots
Broccoli	Commercially grown only in small plots
Head cabbage	Commercially grown only in small plots
Cantaloupe[b]	Poorly adapted. Commercially grown only in small plots
Carrots	Commercially grown only in small plots
Cauliflower	Commercially grown only in small plots
Cucumbers and pickles[b]	Commercially grown only in small plots
Garlic	Poorly adapted. Commercially grown only in small plots
Herbs, fresh cut	Commercially grown only in small plots
Lettuce and romaine	Commercially grown only in small plots
Dry onions	Poorly adapted. Commercially grown
Green onions	Commercially grown only in small plots
Green peas	Commercially grown only in small plots
Hot peppers[b]	Poorly adapted. Commercially grown only in small plots
Sweet peppers[b]	Poorly adapted. Commercially grown only in small plots
Pumpkins[b]	Commercially grown only in small plots
Spinach	Commercially grown only in small plots
Squash[b]	Commercially grown only in small plots
Sweet corn[b]	Commercially grown only in small plots
Tomatoes[b]	Commercially grown only in small plots
Wasabi	Well adapted. Significant commercial production.

[a]Based on compilation by Dr. N.S. Bill Mansour, Extension Specialist—Vegetable Crops, Department of Horticulture, Oregon State University, mansourb@bcc.orst.edu, https://hort.purdue.edu/newcrop/cropmap/oregon/crop/cstl-veg.html

[b]Needs appropriate agronomic techniques

In addition, coastal regions of India are known for evolving and conserving breeds of cattle, buffaloes, goats, and pigs (Kerala), and are rich in marine brackish water and freshwater aquatic biodiversity (fish and shrimp). They have been integrated with several traditional field cropping/coastal agricultural systems to provide sustainable additional income to farmers and livelihood support to local populations and to support the agro-industries with raw material.

Major Challenges

Coastal agriculture benefits from favorable high moisture environmental conditions that can also be supportive of other activities, such as industry and tourism. However, due to its proximity to the ocean/sea, agriculture is threatened by several natural constraints, which are often activated by nature's fury (cyclones, tsunamis, etc.) causing saline air and water; due to its location at the downstream end of river flows by poor quality and insecure supply of water caused by upstream activities; by the limited space for expansion and relocation under severe competition for available coastal land; and by the limited resources in many coastal areas, together with the need to limit potentially harmful effects/impacts of modern agriculture on sensitive and important coastal ecosystems.

Furthermore, the agriculture sector influences, and is influenced by, other sectors. These interactions may be positive, but are more often negative, and revolve around competition for land, water, capital and labor. The negative influences of agriculture on other sectors include pollution of fisheries by agrochemicals and silting of coral reefs and ports resulting from land erosion. In turn, agriculture itself may be negatively influenced by pollution originating from outside the coastal area, or it may induce its own negative impacts, for instance by inappropriate irrigation practices which can lead to the intrusion of saltwater from the sea. The following paragraphs discuss the major challenges faced by coastal agriculture arising due to various factors.

Oceanic/Sea Activities in Coastal Areas and Climate Changes

Flooding and Salinity

Many coastal agricultural areas are flat, while others have steep coastlines. Low-lying agricultural land is frequently subjected to severe drainage and soil salinity problems due to stagnation of rain and runoff water and flooding from rivers or periodic storm surges. Physical damage is caused by windstorms or tidal waves, and sensitivity to airborne salt deposition. The tides induce penetration of seawater far inland in the lower reaches of most coastal water courses during high tides and cause periodic increases in river water salinity that may preclude or complicate its safe use for irrigating agriculture. Low-lying agricultural lands may also be susceptible to shoreline retreat and flooding as a result of coastal erosion. Higher air humidity in coastal areas is favorable to the occurrence and propagation of certain invasive plant diseases and pests that can constrain crop growth and cause health hazards.

Global Warming

Coastal ecosystems are more susceptible to changing global climate. Because of global warming sea levels are rising, due to thermal expansion of seawater, which can cause significant shoreline inundation, overstepping of barrier islands, and loss of intertidal wetlands, leading to increased salinization of coastal embayments and agricultural lands. In addition, other climatic factors, such as increased tropical storm intensity and frequency; changes in precipitation patterns and river flow; changes in seawater temperature range and seasonality; and alteration of coastal currents and upwelling (Bakun 1990; van Geen et al. 1992), affect temperature, nutrient supply and larval transport, whereas modification of intermediate-scale weather patterns that affect winds, currents, and rainfall adversely affect agriculture. Rise in temperature will also adversely affect crop productivity and overall agriculture sustainability in evolving situations.

To overcome these constraints, there is a continued need for screening of crop germplasm in search of new genotypes (genes/alleles) of coastal crops with greater tolerance to salinity and high temperature, for successful cultivation and to enable genetic engineering of the existing cultivars for greater resilience and sustainability. In addition, experimentation may be needed to discover the agronomic modification/diversifications of cultivation practices that can help overcome these constraints.

Shoreline Erosion

Loss or displacement of land along the coastline due to various oceanic or sea activities, which is quite common, may cause physical damage to standing crops and/or loss of agricultural land, reducing the expected output and causing social problems. The principal source of sedimentation along the coastline is from rivers and streams, and thus changes in land use and stream hydrology can affect the supply of sediments to beaches and, consequently, shoreline erosion. Shoreline erosion is also influenced by coastal processes resulting in offshore and alongshore transport of sediments. Therefore, shoreline erosion and hazardous storms greatly affect coastal environment quality (edaphic), limiting agriculture land use and its mitigation.

To restrict or reduce shoreline erosion, artificial or natural physical barriers must be created. Mangrove forests along most coastal lines and raising and maintenance of traditional or improved dams can restrict/limit the forced entry of seawater, reducing shoreline erosion. This will also help in restricting contamination of the freshwater cultivation. In this regard, lessons can be learned from the traditional backwater system that has been developed in the Malabar region of Kerala, India, for management of excess water resulting from sea activities, excessive rain, and so on. Moreover, mitigation, such as management practices, have and can facilitate aquaculture, duck farming, and irrigation of upland crops and their integration.

Developmental Human Intervention

Surface water becomes available to coastal agricultural activities only after it has already been used for upstream activities. Excessive use of water from upstream dams and in irrigation schemes can deprive coastal areas of water for irrigation, and removing silt and regulating floods may affect the fertility of coastal alluvial agricultural land. Conversely, inland encroachment of agriculture onto forested land, slash and burn practices, overgrazing and inappropriate cultivation methods may increase runoff and erosion in catchment areas, with coastal land affected by increased sediment in rivers, lower dry season river flows and increased flooding. The Greater Bengal coastal area has been highly affected by deforestation in the Himalayas several hundred kilometers to the north. The quality of surface water available for coastal agriculture may also be affected by upstream discharges of industrial and urban effluents and by drainage of chemicals and salts from agricultural land into rivers.

Further, the dynamics of alluvial landscapes and natural sedimentation patterns that determine the nutrient and energy flows in coastal areas are being increasingly modified by human activities, particularly by those that affect water flows (dams, increased water extraction, deviation of rivers), and erosion, especially that caused by deforestation. This prevents or slows down vertical accretion, thus aggravating saltwater intrusion and impairing drainage conditions in riverine, delta or estuarine areas. It reduces or blocks sediment supply to the coast itself, which may give rise to the retreat of the coastline through wave erosion. These are additional constraints of coastal areas, which are already prone to natural threats from tidal surges and sea level rise as discussed above.

Space and Resource Constraints Due to Competition and Overexploitation

Coastal areas are restricted to limited extension. Opportunities for expansion or relocation of agricultural activities are, therefore, limited, especially when the coast is bound by mountains. Population growth can lead to pressures on resources such as water and land. This becomes intense with urbanization. As competition for land intensifies, agricultural activities such as grain crops and livestock grazing tend to be marginalized. Smallholders, unable to bear the stress, tend to lose their land and migrate to the coastal towns. The marginalization of low-value agricultural production is affecting peasant agriculture. Increasing pressure on land may lead to overexploitation, with inappropriate land use through the extension of agricultural activities into areas that may not be suitable for agriculture having consequences that may affect coastal ecosystems. Good communication means in coastal areas may facilitate imports of low-cost, high-quality agricultural produce with the consequent loss of outlets for local producers.

Agriculture may be constrained because of the competition for natural resources vis-à-vis other economic activities. Competition for land and water is adversely affecting coastal ecosystems. Agriculture being the major occupier of land for cultivation and grazing may result in habitat loss and loss of biological diversity. Use of coastal wetlands and dunes for grazing in the long term can cause steady land and vegetation degradation.

Therefore, unplanned development of infrastructures, including dams and irrigation schemes to promote agriculture, inland or near the coast, may reduce surface flows, with effects on the timing as well as the overall volume of river flows, and can adversely affect the availability of natural resources, such as land, silt (to contribute to land fertility) and water. Further, irrigation and cropping activities in coastal areas may lower the water table. Irrigation systems extracting groundwater from coastal aquifers may result in exhaustion of these or make them increasingly salinized as saltwater intrudes into them.

Habitat Modification

Infrastructure development for trade, industrialization and human population growth because of rising opportunities in coastal regions are adversely affecting the positive environmental conditions available for agriculture. Development of agriculture increases settlement and population growth, which in turn further affects the local environment. These changes modify the prevailing natural habitat, with habitat destruction adversely affecting the biodiversity, conservation and sustainable use of natural resources. In addition to agriculture, they may also threaten coastal livelihoods, increasing poverty and forcing people to overexploit the remaining resources available to them (e.g. by draining wetlands, destroying forests and overfishing) and change in demography. Therefore, integrated planning is required at the national and international level to facilitate conservation and use of natural resources that are integral to sustainable agriculture.

Toxic Effects of Agriculture

Intensive agricultural activities related to cash crops utilizing modern agriculture technologies with predominant use of inorganic fertilizers, fungicides and pesticides may result in water and land pollution through runoff of agricultural chemicals, and can have a harmful impact/effect agriculture on sensitive and important coastal ecosystems related to marine life. Toxic chemicals and organic wastes in surface water may damage and destroy sensitive coastal ecosystems. Damage to coastal ecosystems has wide-ranging implications for the forestry, fishing, aquaculture and tourism sectors, which may directly rely on them for natural resources. From a wider perspective, agricultural activities can have serious and damaging impacts on coastal ecosystems and on human populations, with health problems arising from pathogens and toxins. Therefore, participatory planning is required involving all stakeholders, including scientists (from concerned disciplines), coastal farmers and so on, for promotion of organic agriculture or innovative technologies with a combination of traditional and modern agricultural technologies.

Introduction of Exotic Species

Many plants and animals that humans introduced benefited from their pre-adaptation to their new environments, including coastal environments, and some became invasive, damaging the biodiversity and agricultural value of the invaded ecosystems. The invasion by non-native organisms is accelerating with human population growth and entrepreneurs looking for new opportunities for greater economic returns and globalization.

Management to remove or reduce the impacts of invasive species should be an ongoing process. An adaptive management approach provides a workable framework that can be readily adopted by stakeholders because it will involve them in applying the best practices and lessons about how to better deal with the problems of animal or plant growth throughout the process. We must decide whether the impacts of an invasive species require rectification or acceptance as a new component of the biota, naturalized and adding to the ecological equivalent of multiculturalism. These are new agri-ecosystems with naturalized plants and animals that have a range of impacts from positive through neutral to negative. As a general rule, positive values and impacts should be encouraged, negative impacts should be discouraged (usually through population suppression and exclusion) and no action is needed for neutral effects.

Ecosystem Perspective

The emerging and widespread environmental threats discussed above pose new challenges to ecosystems and thereby to agriculture. These concerns have not been eliminated, although some of their effects are relatively well understood, and significant advances have been made in their management in several developed countries. Concern is shifting from issues related to single-factor risk assessment to approaches involving multiple-stressor (e.g. combined effects of chemical contaminants and low oxygen) risk assessments and indirect, cascading and scale-related effects on biological resources. Therefore, understanding of such coastal problems requires approaches that focus on ecosystems, populations of naturally growing organisms, and communities of species and the one cultivated/domesticated species. There is now greater concern about the response of ecosystems to the effects of exploitation of resources, nutrient enrichment (as opposed to direct organic loadings), and the indirect effects of human activities on coastal habitats, including man-made habitat, for example agro-ecosystems.

Physical features of coastal ecosystems, such as reefs and belts of mangrove, are important for mitigation of the effects of natural disasters, such as storm-tide surges, shoreline retreat or floods. These features also play an essential role in natural processes, such as land accretion, and help to control coastal erosion and other damages arising from wind and wave action.

The development of initiatives to solve ecosystem-related problems will require more flexibility from the scientific community, universities and funding agencies to promote interdisciplinary science.

Table 2 presents a summary or birds-eye-view of the challenges/constraints faced by coastal agriculture, their causes and possible solutions to overcome, to make agriculture sustainable and more productive both at subsistence and commercial level.

Table 2 Summary of challenges/constraints to coastal agriculture: causes and solutions

Constraints	*Causes*	*Possible control measures/mediation*
Increased flooding, salinity, temperature rise, shoreline erosion	Proximity to sea, climate change, rising temperature and sea level, shoreline erosion	Study of biophysical environment and finding solutions accordingly–efficient irrigation systems, search for tolerant genotypes and use in crop improvement, improved or adoption of appropriate cropping systems
Competition for land, water and other resources due to development	Settlement, population growth, fisheries, aquaculture, forestry, infrastructure and industrial development, tourism	Balanced socio-economic and environmentally friendly development and increased crop productivity per unit area reducing pressure
Habitat modification	Population growth, introduction/development of alternative economic activities	Population control and eco-friendly introduction/development of other sectors
Toxicity and other negative effects due to overcropping, grazing in watershed areas	Impact of agriculture on other sectors and vice versa	Study of current and anticipated negative and positive effects of agriculture and vice versa, practice of organic agriculture and controlled cropping, grazing
Negative biological environment due to species composition	Natural or man-made introduction of exotic/invasive spp.	Physical eradication or integrated biological control of invasive spp.
Ecosystem perspective	Emerging and widespread environmental threats	Understanding the factors followed by interdisciplinary research involving all stakeholders

Need for Planning Integrated Management of Coastal Area, Including Agriculture

Recognizing the fragile nature of coastal ecosystems, the agricultural planning in coastal areas must be in line with national policies on economics, and regional and agricultural development, and these must be consistent with one another. Agricultural plans must include objectives on the efficient use of land and water, the appropriation of new land for agriculture, and the maintenance (or restoration) of the water flows and stocks and water quality necessary to support agriculture and coastal ecosystems, as well as on the use of agrochemicals and other impacting factors.

The objectives of the plan must reflect the description of the prevailing situation and agricultural challenges and opportunities. An integrated approach for management of agriculture in coastal areas may be developed and adopted using the Integrated Coastal Agriculture Management guidelines (FAO 1998) as a basis with a description of development targets, requirements of government interventions and resources; cost-benefit analysis; and provision for regular monitoring with flexibility for on-course corrections. This shall help in carrying out activities in a way that is sensitive of coastal ecosystems.

In order to integrate agricultural planning into overall coastal development, the first stage is to gather relevant and useful information on the environment and resources. This should cover the biophysical and (including edaphic, landscape, biological diversity and cultivated species adapted to, etc.) socio-economic environments (population, gender, literacy, per capita income, etc.), interactions with other sectors (non-agriculture), governance, and the constraints, opportunities and possible alternatives and synergic components for the agricultural sector, such as integration of fisheries and forestry.

The next stage is planning that takes into account the special characteristics of coastal agriculture, ensuring that the plan conforms to overall national objectives for agriculture and honoring the provisions of national legislation and international conventions. During this phase, means to reduce or avoid the negative impacts of agriculture on other sectors should be introduced; these may entail revising provisions of subsidies, taxation and regulations, while introducing specific support services and reviewing the institutional set-up. The outcome may require changes in cropping patterns and cultivation methods to facilitate conservation and effective use of natural resources.

Throughout the process, a participatory approach must be adopted, and all interested parties including farmers, agriculture researchers and other stakeholders such as processors, traders (marketers of agricultural produce) should be consulted and involved. At governmental level, a close link should be maintained with relevant ministries/departments and services dealing with the other sectors. Thus, there should be an attempt to ensure coastal area agricultural development plans which will address the specific characteristics of agriculture in the area, interactions with other sectors and the importance of sustainable practices.

Whatever may be the situation, there are reasons for giving ample attention to agriculture in integrated coastal resource management planning, as follows:

- Agriculture effects on the coastal environment. Sustainable agricultural policies are needed to minimize the negative impacts of inland agriculture on coastal areas and conservation of coastal natural resources.
- Agriculture is mainly concerned with food production. However, in coastal areas, urban growth, the high value of urban goods and services, trade, and the political importance of urban centers marginalize coastal agriculture. Protecting and maintaining peri-urban agriculture in coastal areas can make a substantial contribution to food and nutritional security and employment.
- Agriculture often provides raw materials to industry located in coastal areas and may therefore have considerable economic significance. Fostering the linkage between agriculture and tourism may create demand and supply for food and beverages and thereby employment in the hospitality sector.

Planning for coastal agricultural activities must make explicit allowance for the wide range of farmers' interests and activities, including non-farm activities. The complexity of these factors makes participation or consultation with farmers important and a prerequisite in coastal area planning where it impinges on agricultural activities. Therefore, planning in coastal areas must account for such special characteristics of agriculture, as follows:

- to include a broad range of activities with a variety of products (crops, livestock, poultry, food and industrial raw materials),

enterprises (from smallholdings to large commercial enterprises) and production processes (including food);
- to make a variety of contributions to the local economy, social cohesion and the maintenance of the cultural traditions of a society;
- it is dependent on natural growth processes and, to a greater or lesser extent, on land and its associated natural resources and environment.

Small-scale agriculture must take the special requirements of gender equality into account (Barbier 1989).

Other components that can be included in broad planning for coastal environment management are promotion and maintenance of mangrove forests and/or development and maintenance of traditional dams along most coastal lines to restrict intrusion of seawater, contaminating freshwater cultivation. In this regard, lessons can be learned from traditional systems such as the creation of a network of backwater channels in the areas suffering from surplus water caused by excessive rains and regular action of sea waves and shore currents in the coastal areas, and in the areas below sea level. The backwaters are a chain of brackish lagoons connected by natural and artificial canals. They represent a unique ecosystem, where freshwater from the rivers meets the seawater. In certain areas of Kerala such as the Vembanad Kayal, a barrage has been built (Kumarakom) to restrict the entry of saltwater from the Arabian Sea to deep inside the mainland, keeping the freshwater intact. Such freshwater is extensively used for irrigation of crops. The backwater channels have also been used for transport, and a large amount of local trade in the region is carried by inland navigation. Fishing, fish curing and wet agriculture are the other important activities that have been performed by the local people in the backwaters for centuries. Encouraged by this unique traditional agroecosystem, agricultural efforts have been strengthened in recent times with the reclamation of more lands for rice cultivation, particularly in the Kuttanad area of Kerala.

Further, reconciling with such a landscape, the local communities have also been able to convert the situation from constraints to opportunities with initiation of entrepreneurships for boat making and the coir industry and other traditional crafts in the backwaters for livelihood support. Development of the backwater system in low-lying areas also facilitates aquaculture, duck farming, and their integration with crop farming and irrigation of upland crops. The traditional below sea level rice cultivation system on the Malabar coast has been recognized by the FAO (https://

www.fao.org/giahs/giahsaroundtheworld/designated-sites/.../kuttanad...system/en/) as one of the most globally important agriculture systems.

The plans may also include large-scale screening of germplasm of coastal crops (rice) against the various stresses/constraints caused, such as submergence, salinity, tolerance and high temperature, to facilitate breeding of cultivars with greater resilience against these stresses of coastal agriculture and to improve economic gains. Introduction of exotic germplasm of the crops indigenous or adapted to coastal areas, and exotic crops, such as oil palm, avocado and rambutan, in the Indian context would help in genetic improvement of the coastal crops, developing new cropping systems and thereby the sustainability of productivity of both subsistence and commercial agriculture.

Conclusion

Moist and fertile soil conditions of coastal areas often provide a favorable environment for agriculture, which provides food and support to agro-based industries and other sectors. However, temporal ocean/sea activities and competition for natural resources (land and water) and interactions (both negative and positive) with other sectors are on the rise, producing future challenges for coastal agriculture. Therefore, to meet these challenges an effective management and development plan for coastal areas needs to be developed integrating sustainable agriculture, concentrating on reducing and/or ensuring control of negative effects on agriculture and other sectors and conserving natural resources. This would require participatory planning and coordinated allocation of resources to different activities to minimize negative effects and maximize the contribution of various sectors in sustainable development. Integrated coastal agriculture management planning would require information on resources, effects of different ocean/sea activities and interactions among various sectors and their development over time, and identification of ways to control or limit the harmful effects of competitive and antagonistic interactions, encouraging benefits from complementary interactions.

Acknowledgement The author is grateful to Drs. S N Nigam and S P S Beniwal, former colleagues from International Crops Research Institute for the Semi-Arid Tropics, Patancheru, Telangana, India, for their critical review of the manuscript and improving it with valuable suggestions.

References

Bacow, L. S., & Wheeler, M. (1984). *Environmental Dispute Resolution.* (pp. 372). New York: Plenum Press.

Bakun, A. (1990). Global Climate Change and Intensification of Coastal Ocean Upwelling. *Science, 24,* 198–201. https://doi.org/10.1126/science.247.4939.

Balasundaram, C., Dheepa, A., & Mariappan, P. (1999). Fish Diversity in Grand Anicut, River Cauvery (Tiruchirapalli, Tamil Nadu). *Zoos' print, 14*(8), 87–88.

Barbier, E.B. (1989). *The Economic Value of Ecosystems.* 1. *Tropical wetlands.* (GK89–02) (pp. 12). Gatekeeper Series. London: Environmental Economic Center.

Barbier, E. B. (1990). Alternative Approaches to Economic-Environmental Interactions. *Ecological Economics, 2,* 7–26.

Barbier, E. B. (1992). *The Nature of Economic Instruments: A Brief Overview.* (GK92-02) (pp. 9). Gatekeeper Series. London: London Environmental Economic Center.

Biswas, B., Ghosh, D. C., Dasgupta, M. K., Trivedi, N., Timsina, J., & Dobermann, A. (2006). Integrated Assessment of Cropping Systems in the Eastern Indo-Gangetic Plains. *Field Crops Research, 99,* 35–47.

Deb, D. (2000). *Folk Rice Varieties of West Bengal: Agronomic and Morphological Characteristics.* New Delhi: Vrihi Research Foundation for Science Technology & Ecology.

FAO Guidelines. (1998). *Part B: Integration of Agriculture into Coastal Area Management in: Integrated Coastal Area Management and Agriculture, Forestry and Fisheries.* Rome: Food and Agriculture Organization of the United Nations.

FAO. (2008). Conservation and Adaptive Management of Globally Important Agricultural Heritage Systems. PIMS 2050 (pp. 37). Terminal Report, Project Symbol: UNTS/GLO/002/GEF Project ID: 137561 Rome: Food and Agriculture Organization of the United Nations. www.fao.org/fileadmin/.../giahs/PDF/GIAHS_B_terminalReport.pdf 2008

Jayan, P. R., & Sathyanathan, N. (2010). Overview of Farming Practices in the Water-Logged Areas of Kerala, India. *International Journal Agricultural and Biological Engineering, 3*(4), 1–43.

Jayaram, K. C., Venkateswarlu, T., & Ragunathan, M. B. (1982). *A Survey of the Cauvery River System with a Major Account of its Fish Fauna.* Occasional paper no.36. (pp. 115). Calcutta: Zoological Survey of India, Ministry of Environment & Forests, Government of India.

Leena Kumari, S. (2012). Status Paper on Rice in Kerala (pp. 32) – Rice Knowledge Management Portal. Directorate of Rice Research, Rajendranagar, Hyderabad, India. (Status Paper on Rice in Kerala.pdf.).

Ministry of Statistics and Programme Implementation. (2020). Quarterly Bulletin Periodic Labour Force Survey (PLFS) (APRIL–JUNE 2019). Available at http://www.mospi.gov.in. Accessed 2 Feb 2020.

Raman, K. V. (2008). Irrigation in South India (up to 1300 AD): Techniques and Management. In D. P. Chattopadhyaya (G. Ed.), Vol. V. Part I. *History of Agriculture in India (up to c.1200 AD)*, Lallanji Gopal & V. C. Srivastava (Eds.), *History of Science, Philosophy and Culture in Indian Civilization* (pp. 496–505). New Delhi: PHISPC Centre for Studies in Civilization.

Salam, M. A., Sathees Babu, K., & Mohanakumaran, N. (2008). Home Garden Agriculture in Kerala Revisited. Accessed at: https://www.unu.edu/unupress/food/8F163e/8F163E07.htm

Sasidharan, N. K., Abraham, C. T., & Rajendran, C. G. (2012). Spatial and Temporal Integration of Rice, Fish, and Prawn in the Coastal Wetlands of Central Kerala, India. *Journal of Tropical Agriculture, 50*(1–2), 15–23.

Shehana, R. S., Sathees Baby, K., & Abdul Salam, M. (1992). Spices: A Multipurpose Homestead Component in South Kerala. *Spice India, 5*(9), 15–18.

Siddiq, E. A., Saxena, S., & Malik, S. S. (2006). Rice. In B. S. Dhillon, S. Saxena, A. Agrawal, & R. K. Tyagi (Eds.), *Plant Genetic Resources: Food Grain Crops* (pp. 27–57). New Delhi: Narosa Publishing House.

Singh, A. K. (2012). Probable Agricultural Biodiversity Heritage Sites in India: XIII. Lower Gangetic Plain or Delta Region. *Asian Agri-History, 16*(3), 237–260.

Singh, A. K. (2013). Probable Agricultural Biodiversity Heritage Sites in India: XVIII. The Cauvery Region. *Asian Agri-History, 17*(4), 353–376.

Singh, A. K. (2014a). Probable Agricultural Biodiversity Heritage Sites in India: XX. The Konkan Region. *Asian Agri-History, 18*(3), 257–282.

Singh, A. K. (2014b). Probable Agricultural Biodiversity Heritage Sites in India: XXI. The Malabar Region. *Asian Agri-History, 18*(4), 311–341.

Singh, A. K. (2015). *Agricultural Biodiversity Heritage Sites and Systems in India* (pp. 467). Secunderabad: Asian Agri-History Research Foundation. ISBN 81-903963-4-X.

Torane, S. R. (2009). *An Econometric Analysis of Farming Systems in North Konkan Region of Maharashtra*. Ph.D. Thesis, University of Agricultural Sciences, Dharwad, Karnataka, India.

Vanaja, T. (2013). Kaipad – A Unique, Naturally Organic, Saline Prone Rice Ecosystem of Kerala, India. *American Journal of Environmental Protection, 2*(2), 42–46. https://doi.org/10.11648/j.ajep.20130202.12.

Van Geen, A., Luoma, S. N., Fuller, C. C., Anima, R., Clifton, H. E., & Trumbore, S. (1992). Evidence from cd/Ca Ratios in Foraminifera for Greater Upwelling off California 4,000 Years Ago. *Nature, 358*, 54–56.

Anurudh K. Singh is former head, Division of Germplasm Conservation, National Bureau of Plant Genetic Resources, ICAR, New Delhi, India, and Cytogeneticist and Germplasm Specialist, ICRISAT, Patancheru, India.

Land Mismanagement and Coastal Disasters

Gaurika Chugh

The Problem

India has a long coastline of 7516.6 kms, out of which the mainland and island territories consists of 5422.6 kms and 2094 kms respectively. The extensive coastline covers a total of nine states and two union territories[1] along the west and east coast of the country, inhabited by 560 million and 0.44 million in coastal states and island territories, respectively. The massive coastline of the country is subjected to rapid changes due to natural and anthropogenic factors. The natural changes in the form of waves, tides, winds, near-shore currents, storms, sea-level rise are further exacerbated by anthropogenic interventions due to the destruction of mangroves, building up of coastal structures such as groins and jetties, building up of dams, hardening of shorelines with seawalls and offshore dredging. Furthermore, the pristine and marine ecosystem of the coastal land has been overhauled by the rampant urbanization and growing numbers of large towns and cities along the coast, sand mining, deforestation, monoculture and development of tourism industry. These developmental activities along the coast have indeed resulted in a major overhaul of the land-use change and is the cause for impending coastal disasters.

G. Chugh (✉)
Jawaharlal Nehru University, New Delhi, India

A. Singh et al. (eds.), *Development in Coastal Zones and Disaster Management*, Disaster Research and Management Series on the Global South, https://doi.org/10.1007/978-981-15-4294-7_6

Conserving Fragile Coastal Land

The National Centre for Coastal Research (NCCR) in its report submitted to the Ministry of Earth Sciences[2] (July, 2018) has revealed that between the period 1990 and 2016, approximately 34 per cent of the Indian coastline (2246.49 km) has been eroded and this has been primarily due to anthropogenic factors which have augmented the natural process of coastal change. As per the report submitted by NCCR, it has estimated that along the west Coast, Kerala is subjected to maximum coastal erosion of 263.04 km (45 per cent), followed by Maharashtra which has 178.26 km (24 per cent) of the total coastal loss. Along the east coast, it is West Bengal which has resulted in a loss of 336.52 km (63 per cent) of coastline, followed by Puducherry which has lost 23.80 km (57 per cent) of coastline. The depleting coastline has been mainly due to growing urbanization, reclamation of low-lying land by private developers, construction on river-beds and by real estate companies for the building up of recreational centres, resorts and residential complexes.

The depleting coastal land and its ecosystem led to the enactment of Coastal Regulation Zone (CRZ) Notification 1991 and it employed certain land-use restrictions arounds coasts, lakes, wetlands, rivers and other water bodies. The CRZ 1991 directed the coastal states and Union Territories to prepare Coastal Zone Management Plan (CZMP) within one year from the implementation of the notification. As per clause (d) of Sub-rule 3 of the Rule 5 of the Environmental Protection Rules 1986, the entire coastal stretch, which was subjected to tidal action (in the landward side) up to 500 metres from the High Tide Line and the land between Low Tide Line (LTL) and (High Tide Line) would be referred to as CRZ. The CRZ 1991 notification classified the coastal area into different zones and provided rules for regulation of development activity for these four zones, namely CRZ-I (prohibits new construction within 500 metres of HTL and between LTL and HTL), CRZ-II (prohibits construction or reconstruction of buildings), CRZ-III (area upto 200 metres from the HTL is to designated as "No Development Zone", with the exception of repairs of existing authorised structures) and CRZ-IV (coastal area in the Andaman and Nicobar, Lakshadweep islands). This notification aimed to protect the fragile coastal zones from unplanned developmental activity. This was soon challenged by the 1994 notification, which reduced the distance from 100 m of HTL to 50 m and also exempted construction and development activities from the 200 m "No Development Zone" (NDZ). These amendments were made after the report submitted by the Vohra Committee to the central government. This amendment to the CRZ notification led to massive outcry among the environmentalists and was soon struck down by the Supreme Court in the *Indian Council of Enviro-Legal Action v Union of India and Others* (1996) case. The apex court in this case[3] held that *"both development and environment must go hand in hand ... in order to prevent ecological imbalance and degradation that development activity is sought to be regulated"*. The non-enforcement of CRZ 1991 notification was taken to the Supreme Court and it directed all the states and UTs falling under the coastal zone to implement and enforce the CZMP.

The enactment of CRZ notification was soon forgotten and there was widespread destruction of fragile coastal ecology by superseding the private interests over the needs to protect the ecosystem. The need to protect the fragile ecology should not be parochial and short-sighted but should devise the implementation of laws in such a manner so as to achieve intra and inter-generational equity. This was highlighted in the *K.R. Ratheesh & Ors. v. State of Kerala*[4] case, in which, there was large-scale land reclamation due to coming up of private resorts along the Nediyathuruthu and Vettilathuruthu islands which lay on the ecologically fragile Vembanad lake, which is the longest and second largest wetland in the country and supports immense aquatic biodiversity and is abode to a large number of migrating birds. Therefore, construction by the Vamika resorts on Vettilathuruthu island without getting environment clearance from the Ministry of Environment and Forests (MoEF) stands in clear violation of the CRZ 1991 and 2011 notifications which declared the Vembanad backwaters as Critically Vulnerable Coastal Area (CVCA). The court ordered the demolition of the construction on the lake as it stood in clear violation of the CRZ notification. Justice K. M. Joseph held that,

> "*the Court when it enforces laws in protection of the environment becomes the custodian of the interests of not only the present generation, but also of posterity.......prompt and effective enforcement of such laws will in due course result in the inculcation of the values embedded in the law as habits which will be followed by the future generations*". This laid the roadmap to land use restrictions in ecologically fragile regions which is indeed crucial for sustainable

development for the years to come. (K. R. Ratheesh & Ors v. State of Kerala 2011 W.P © Nos. 19564/11 U)

Similarly, in *Antony A.V. v. Corporation of Cochin*,[5] the single Judge Bench led by Justice A. V. Ramakrishna Pillai in 2014 went a step forward and ordered the demolition of multi-storied DLF residential complex on the eastern bank of Chilavannurkayal backwaters in Kerala and held that,

> *Authorities are constituted under various enactments to see that the environment is protected and to see that the present topography which keeps the ecological balance is not disturbed. The purpose of these laws is to preserve nature for posterity. If the violation of these laws are allowed to become the order of the day, the existence of life would be at peril. Right to life guaranteed by our Constitution takes in innumerable rights, including the right to enjoy nature in the present form. Indiscriminate invasion of nature to the detriment of others is an invasion of right to life. Nature which is the property of the nation cannot be allowed to be scrambled by a minority violating all laws.* (Antony A.V. v. Corporation of Cochin WP © No. 27248 of 2012 (E))

This judgement laid the foreground that sustainable land management is indispensable for any form of development and linked it up with the fundamental right to enjoy nature which cannot be tampered by any administrative authority. However, the judgement was short-lived and was soon challenged by the divisional bench which ordered a stay on the demolition of the structure and rather imposed a fine of Rs. 1 crore on the private developer. The case was eventually taken up by the Supreme Court in *Kerala State Coastal Zone Management Authority (KCZMA) v. DLF Universal Ltd*[6] (2018), which laid heavily on the lackadaisical approach of the various administrative authorities and even snapped the KCZMA of *"not having a an afternoon siesta but a kumbhakarna sleep albeit of almost four years"*. The Centre for Earth Sciences (CESS) in its report submitted in 2011 has said that most of the construction has taken place on the low-lying areas of filtration ponds and paddy fields which has actually resulted in the pushing back of the backwaters by almost five metres. However, the authorities woke up much after the construction had taken place and the lackadaisical approach of the authorities led the apex court to dismiss the petition filed by KCZMA which was filed to quash the order of the divisional court which had put a stay on the demolition of the construction. It cannot be denied that the administrative authorities were really callous and careless in dealing with the protection of fragile ecology but the decision of the apex court to decide the case in favour of DLF was really unfortunate.

Conserving Fragile Coastal Land

Efforts to conserve the fragile coastal land and ensure sustainable development have significantly been taken up at both the state and community level. At the state level, to protect the fragile Western Ghats from the wrath of expanding linear development, various committees like Pronab Sen (2000), Gadgil (2011) and Kasturirangan (2013) were set up to suggest measures to protect the fragile ecosystem of the coastland. However, the recommendations suggested by these committees found little support and later on succumbed to the populist resistance to enforce development in lieu of protecting the ecosystem.

The Western Ghats spanning an area of 1600 kms across six states,[7] which is an abode to the pristine and marine ecosystem and holds various endemic plant and animal species, is facing an extreme backlash due to the changing land-use pattern, illegal, rampant mining and quarrying on fragile land, urbanization, construction of dams, setting up of hydroelectric power stations, loss of mangroves, deforestation and construction in catchment areas, mechanized fishing and growing tourism industry along the coast. Traditionally, the land-use pattern along the western coast has been mainly paddy cultivation in the valleys whereas the cultivation of millets and legumes mainly predominated the hill slopes. Such pattern of land use has been completely shifted since the colonial times to the intensification of commercial crops such as tea, coffee, cardamom, rubber, pineapple and timber plantations. This was supplemented with the establishment of community boards to expand their production and cultivation. Such linear expansion of commercial crops has led to a major overhaul in the land-use pattern along the coasts and has led to changes in land use from forests to agricultural use and has indeed resulted in fragmentation of forests, soil erosion, pollution of rivers and degradation of ecosystem. The pressure on land has also resulted in the felling of natural forests and replacing it with the plantation of exotic species such as eucalyptus and Acacia auriculiformis (WGEEP Report 2011, p. 31). Recent times have also seen the development of a large number of growing towns and cities along the coast, population pressure and growing tourism industry along the coasts which has resulted in the displacement of indigenous and local communities from the periphery to the mainland.

In view of the depleting ecosystem of the Western Ghats and changing land-use pattern, MoEF constituted a Western Ghats Ecology Expert Panel (WGEEP) under the chairmanship of Prof. Madhav Gadgil to suggest measures to revamp the diminishing coasts. The Gadgil Commission submitted its report in 2011 to the MoEF and after a careful examination of the fragile coastal land and its diminishing ecosystem, designated the entire Western Ghats as an Ecologically Sensitive Area (ESA) and demarcated the entire region into three Ecologically Sensitive Zones (ESZ) zones namely, ESZ 1, ESZ 2 and ESZ 3. The methodology adopted by the Gadgil commission to demarcate the region into three zones was based on the severity to preserve and conserve the coasts and was a follow-up to the criteria based on endemism as suggested by the Pronab Sen Committee, which was set up in 2000 by the MoEF. Sustainable land management in coastal areas constitutes a significant part of the Gadgil Commission which has suggested that the entire Western Ghats falls under ESA and therefore steps should be taken to protect entire fragile area and there should be a complete ban on change in land-use from forest to non-forest use or agricultural to non-agricultural use and there should be a complete ban on the use of mining and quarrying activities in ESZ 1 and ESZ 2. The Commission also recommended the setting up of a statutory authority known as Western Ghats Ecology Authority (WGEA), in joint collaboration with the centre and the states that will oversee the overall management and planning of developmental activities in the three ecologically sensitive zones.

In a follow-up to WGEEP, the MoEF constituted another High-Level Working Group (HLWG) under the stewardship of Dr. K. Kasturirangan in 2013 to examine the recommendations suggested by the Gadgil Committee on Western Ghats. Instead of demarcating the entire Western Ghats region as ESA as per the Gadgil Committee, the Kasturirangan committee suggested that only 37% of the Western Ghats falls under the purview of Ecological Sensitive Zone (ESZ). The committee (2013) examined the entire Western Ghats region and came to a conclusion that "*out of the total area of Western Ghats which covers an area of 1,64,280 km^2, approximately 60 per cent of the area falls under the cultural landscape which is dominated by human settlements, agriculture and plantations and about 41 per cent of the remaining land area can be further classified as natural landscape*". Out of this 41% demarcated area as natural landscape, about 37% can be classified as ecologically fragile and needs to be protected from expanding developmental activities, urbanization, mining and quarrying activities.

Apart from the formation of various committees that primarily remained unattended due to the escalation of populist resistance to development, the judiciary has played a major role in protecting the fragile ecosystem. Judicial activism in the form of filing Public Interest Litigations (PILs) by various NGOs, public spirited citizens or by community-based organizations like Nagrik Samitis has played a major role in dismantling the edifice of environmental degradation. In 1987, Goa Foundation, an environmental action group led by Norma and Claude Alvares filed a first-ever PIL in the Supreme Court to save the pristine coastal ecosystem from the wrath of sand mining. The protection of sand dunes is indeed crucial to preserve the ecosystem as they act as buffers to protect the hinterland from the ravages of the sea. They are also significant to protect the land from inundation and also act as an important source to recharge the fresh water aquifers. The change in land use along coasts has mainly been due to the diversion of forest land for building up of tourist spots or giving away land for mining. As per the Forest Conservation Act, 1980, the use of forest land for non-forest purposes requires permission from the MoEF. A series of PILs has been filed which have challenged the conversion of forest land, and this laxity has been mainly due to faulty land records (Alvares 2010, p. 7). The land records are not regularly updated and this becomes a bone of contention for giving away forest land for non-forest use. For instance, in *Tree Officer v/s Salgaoncar & others*, an industrialist had sought permission to fell off trees and as per the records of rights (RoR), the entire land had been demarcated as barren land; however, on inspection it was found that the entire area was forest land (Alvares 2010, p. 7). The court ruled that faulty land records cannot be taken as a criterion for demarcating the entire land as forest area.

Environmental degradation mainly affects the people living on the periphery and who are mostly depended on the ecosystem for their sustenance. Large-scale mining, conversion of forest into non-forest land and building of dams has led to mass-scale displacement of indigenous and tribal communities. These indigenous communities have been termed by Gadgil and Guha as 'ecological refugees who live on the margins of islands of prosperity' (2013, p. 4). The trade-off between protecting the ecosystem and development has led to the intensification of community resilience among the traditional and indigenous communities. Various community-based organizations, civil societies, environmentalists and people from all walks of live have altogether come up in the forming of various community-based movements like Silent Valley Movement in Kerala (1978), Appiko Movement in Karnataka (1983), Chilika Bachao Andolan in Odisha (1990s) and Narmada Bachao Andolan (1989) led by National Alliance for People's Movements (NAPM).

The depleting land use in coastal areas is also augmented with the spike in the mechanization of fisheries by permitting fishing in waters further off the coast. While the earlier system of fishing along the coast by artisanal fisherfolk was mainly for the use of self-sustenance using small rowing and sailing boats, the growing mechanization of fishing industry has led to the inception of trawlers and purse seiners and growing of cold storage and canning facilities and escalation in the export of marine products (Gadgil and Guha 2013, p. 17). The growing commercial industries along the coast has indeed resulted in the movement of fish away from the coast and mechanization has led to overfishing and a decline in total fish catch in several states. This has indeed resulted in a loss of livelihood for communities who were dependant on the coasts for their survival and the mechanization along the coast has resulted in depletion of the ecosystem. This resulted in the inception of fisherfolk movement in Kerala with the formation of cooperatives and was led by Catholic priests in 1960s who first began to organize the fisherfolk communities into welfare cooperatives. Their first unified organization was constituted under the leadership of Fr. Paul Arrakal of Allapuzha and Fr. Albeit Parsiavila of the Kallam union in 1977 and was named as Kerala Lateen Catholica Matsya Thozhilali Federation (KLCMTF). To give the organization a more holistic and secular character, it was renamed as Kerala Swathantra Matsya Thozhilali Federation (KSMTF). It was in 1989 that the National Fisherworkers Forum (NFF), in collaboration with the KSMTF, organized a pan-India march which was known as Kanyakumari march to save coasts that started from Calcutta in the Eastern Coast, crossing Bombay on the West coast and finally reaching Kanyakumari. Their slogan was 'protect waters, protect live' and it was a first-ever kind of movement that brought in the livelihood concerns of communities dependant on coasts by linking it to the demand for an environmentally sustainable development model.

The Real Picture Post Kerala Floods and Its Impact on Coastal Ecosystem

Notwithstanding the recommendations suggested by the Gadgil (2011) and Kasturirangan (2013) reports that sought certain land-use restrictions for the ecologically fragile Western Ghats, the torrential rains that swept the Kerala coasts from 15 to 17 August turned the deluge into an unprecedented disaster. The catastrophic disaster was a result of unplanned developmental activities that was being carried out along the coast in the form of deforestation, land reclamation on river beds, illegal mining, urbanization and development of tourist hotspots. The Western Ghats Ecology Expert Panel Report (WGEEP) in 2011 had recommended the demarcation of the entire Western Ghats as Ecologically Sensitive Area (ESZ) to protect it from the expanding developmental activities. However, the measures suggested by the Gadgil Committee (2011) and Kasturirangan Committee (2013) were left unattended and this has been a major cause for the impending coastal disasters. Even the CRZ notification of 1991 and 2011 had classified the entire area into four zones and imposed certain land-use restrictions along the coasts but these were circumvented at the cost of protecting the ecology of fragile regions. Singh (2016) had already stated that "*any wavering on implementing the CRZ law will make areas closer to water bodies vulnerable to disasters amounting to huge socio-economic destruction and loss of lives*" (p. 70).

The havoc caused by the unprecedented floods resulted in the destruction of state's wealth and its ecosystem. It has been estimated that the floods resulted in a total loss of 433 human lives, affected a staggering of 5.4 million people and displaced a total of 1.4 million people, resulted in a total loss of Rs. 26, 720 to the state exchequer and the total recovery needs post-disaster are estimated to be Rs. 31,000 crores.[8] As far as floods are concerned in Kerala, it is more than a usual activity that happens every year for the communities inhabiting the low-lying areas of Kuttunad and Alleppy. However, the hazard turned into a disaster mainly due to two reasons: first, the change in land-use pattern in the form of illegal construction on river beds and change in traditional agricultural production and unauthorized stone quarrying along the coast resulted in the destruction of its ecosystem; and second, the laxity, rapacity and unpreparedness of the administrative authorities and their lack of coordination with the local self-government. This has been highlighted in the report[9] by Advocate Jacob P. Alex who was appointed by the Kerala High Court as

Amicus Curie to dispose of a number of PILs that had flooded the court post Kerala floods. The report clearly underpins that the losses that were accrued were a result of the mismanagement by dam authorities, lack of preparedness at the state level for ensuring flood management and changing land-use pattern along the river beds that has led to encroachment of flood plains. It has been brought to the fore that none of the 79 dams in Kerala were used for the purpose of flood control despite the regulation as stipulated by the National Water Policy (2002), National Disaster Management Authority (NDMA) guidelines on flood management and Real Time Integrated Operation of Reservoirs (RTIOR). As per these guidelines, it is said that water should be released before it reaches the mark of Full Reservoir Level (FRL) and it should never be allowed to cross the mark of Maximum Storage Level (MSL). However, the administrative authorities were quite lackadaisical in their approach and water was released only when it reached the FRL mark and was close to MSL.

The findings of the Amicus Curie report was reiterated much earlier by the Special Centre for Disaster Research (SCDR) report[10] (2018) on Kerala Floods which highlighted the ecosystem degradation on account of unplanned development activity and lack of preparedness and coordination among Kerala State Disaster Management Authority (KSDMA), dam authority, Kerala State Electricity Board (KSEB) and their neglect of panchayat institutions. The changing coastal ecosystem with unplanned development activity over river-beds, mining and quarrying along the coasts, encroachment of low-lying areas and the changing agricultural land use are the reasons that had gone into the making of the Kerala disaster. Out of the total coastline length of 592 km of Kerala, it has been estimated that between 1990 and 2016, about 263.04 km or 45 per cent of the coastal land has been eroded (*Report by NCCR on National Assessment of Shoreline changes along Indian Coast: Status Report for 26 years, July 2018*). The ebb and flow of coasts is a natural course, but such devastation of the coastline has been mainly due to anthropogenic reasons which have changed the land-use pattern of the state and thus have been the cause of unprecedented damage during the Kerala floods. In the name of eco-tourism, a large part of hill slopes and mangroves have been destroyed and privatized to set up resorts, rejuvenation centres and private mansions. Conservation of mangroves is crucial for sustaining the coastal ecosystem as they are natural wetlands and act as a bulwark against the catastrophic effect of disasters. It has been estimated that this pristine and primordial forest resource has reached the lowest ebb of just 25 sq. km at present, down from 700 sq. km in 1957 (Singh et.al. 2018, p. 15).

It has been estimated that the floods caused damage to 45,000 hectare of farm fields (https://www.hindustantimes.com/india-news/kerala-floods-damage-45-000-ha-of-farfields-paddy-banana-spices-among-worst-hit/story-bF0Iully9ppN6vWnuDcGHL.html). The traditional agricultural production in Kerala, which was based on paddy cultivation, has been shifted to the commercialization of crops such as tea, coffee, rubber, cardamom and timber plantations. Paddy cultivation acted as a natural ecosystem to conserve the wetlands and also helped to recharge the groundwater aquifers. This was also accompanied by land encroachment in the paddy fields by the private developers which has resulted in a major overhaul of land-use change. As per the data provided by the Department of Economics and Statistics, Government of Kerala, the state witnessed a drastic decline in paddy cultivation, from 48% in 1967–68, to 15% in 1995–96 and finally settled at a meagre 10% per cent of the total crop area in 2004–05 (http://www.ecostat.kerala.gov.in/images/pdf/publications/Agriculture/data/old/paddy_c_9596_0405.pdf). The area under rubber plantation grew from 1.0 lakh hectares in 1957–58 to 5.48 lakh hectares in 2013–14 and around 14 per cent of the total geographic area of Kerala is under rubber cultivation (Chattopadhyay 2015, p. 10).

Land mismanagement in the form of illegal construction along the low-lying areas and flouting the norms of the CRZ 1991 and 2011 notification have been the cause of the major devastation caused due to Kerala floods. This was remarked in the recent case[11] in the Supreme Court in which a bench of Justices Arun Kumar and Navin Sinha ordered the demolition of five housing complexes in Maradu municipality in Ernakulam that were constructed by flouting the norms of Coastal Zone Management Authority (CZMA). The bench took cognizance of the land mismanagement and recent wreckage caused due to Kerala floods and held that "*we take judicial notice of recent devastation in Kerala which had taken place due to heavy rains compounded by such unbridled construction activities resulting in colossal loss of human life and property due to such unauthorised activity*".

Conclusion

The recent devastation caused by the Kerala floods lead us to introspect what went wrong, how and why? The havoc created by the Kerala floods did not just end with it, as disasters have a cascading effect. Several districts of Kerala were engulfed in a drought-like situation post the floods. As large volumes of water was released from the upper echelons of Idukki and

Cheruthony dam, it led to the destruction of the ground water aquifers. There has been a massive depletion of the ground water table and the top soil has been eroded. It can be concluded that land-use restrictions are indispensable to sustainable development. However, the law stands mighty and robust in terms of protecting the fragile and pristine ecology of coasts; in fact, what makes it weak and feeble is the roughness, callousness and lackadaisical approach of the authorities who are entrusted with the duty to implement the law. Moreover, the existence of too many authorities like KSDMA, KCZMA, KSEB, MoEF and local-self-governments work in silos with little or no coordination between them; this makes the implementation of due process of law cumbersome. It is therefore necessary that the administrative authorities who are in charge of implementing the land-use restrictions and protecting the ecosystem should be the forebearers of implementing the due course of law and there should be strict accountability on the part of those who flout the norms that are devised to protect the ecosystem.

Notes

1. The states and union territories covering the coastline are: Gujarat, Maharashtra, Goa, Karnataka, Kerala, Tamil Nadu, Andhra Pradesh, Odisha, West Bengal, Daman & Diu and Puducherry. The island territories along the coast are Andaman & Nicobar Islands (East Coast, Bay of Bengal) and Lakshadweep (West Coast, Arabian Sea).
2. The National Centre for Coastal Research (NCCR) has submitted a report in July 2018 to the Ministry Of Earth Sciences titled *"National Assessment of Shoreline changes along Indian Coast: Status Report for 26 years"*. The NCCR has carried out an extensive study on the changes accrued on the coastline along the nine states and two union territories.
3. Indian Council for Enviro-Legal Action v. Union of India (1996) 5 SCC 281.
4. K.R. Ratheesh & Ors v. State of Kerala 2011 W.P © Nos. 19564/11 U.
5. Antony A.V. v. Corporation of Cochin WP ©. No.27248 of 2012 (E).
6. Kerala State Coastal Zone Management v. DLF Universal Limited (2018) 2 SCC 203.
7. The Western Ghats cover an area of 1600 kms or 1.60 lac sq. kms across six states, Gujarat, Maharashtra, Goa, Karnataka, Kerala and Tamil Nadu.
8. Kerala Post Disaster Needs Assessment Report on Floods and Landslides, 2018 (PDNA). This report is jointly prepared by the United Nations, Asian Development Bank, World Bank and Government of Kerala.
9. Amicus Curie in W.P. No. 2651 of 2019.

10. '2018 Kerala Floods: A Report on Governance and Legal Compliance', New Delhi: UPE2, *SCDR-NIDM Research Publications.*
11. CIVIL APPEAL NOS.4784–4785 OF 2019 (Arising out of SLP (C) Nos.4227–4228 of 2016).

References

Alvares, N. (2010). *Political Struggle Through Law the Public Interest Litigation (PIL) Routeto Environmental Security in India with Special Reference to the Environment Movement in Goa.* WGEEP Commissioned Paper. http://www.westernghatsindia.org/commissioned-papers

Antony A V v Corporation of Cochin. (2012). WP(C), No 27248 of 2012 (E).

Gadgil, M., & Guha, R. (2013). *Ecology and Equity. The Use and Abuse of Nature in Contemporary India.* Routledge.

Indian Council of Enviro-Legal Action v Union of India and Others. (1993). Writ Petition (WP)(C) No 664 of 1993, 18 April 1996.

Kerala Post-Disaster Needs Assessment Report on Floods & Landslides. (2018). Report Prepared Jointly by United Nations, Asian Development Bank, World Bank & Government of Kerala.

Kerala State Coastal Zone Management v. DLF Universal Limited. (2018). 2 SCC 203.

National Assessment of Shoreline Changes Along Indian Coast: Status Report for 26 Years 1990–2016. (2018, July). Report Submitted by National Centre for Coastal Research (NCCR) to Ministry of Earth Sciences, Government of India.

Ratheesh, K.R., and Others v State of Kerala and Others. (2011). WP(C), No 19564/11 Decided on 25 July 2013.

Report of the Amicus Curie in W.P. No. 2651 of 2019.

Report of the High Level Working Group on Western Ghats. (2013). Submitted to the Ministry of Environment and Forests, Government of India.

Report of the Western Ghats Ecology Expert Panel. (2011). Submitted to the Ministry of Environment and Forests, Government of India.

Singh, A. (2016). Coastal Ballads and Conservation Ironic. *Economic and Political Weekly, LI*(7), 70–75.

Singh, A., et al. (2018). *2018 Kerala Floods: A Report on Governance and Legal Compliance.* New Delhi: UPE2, *SCDR-NIDM Research Publications.*

The Kerala State Coastal Zone Management Authority v. the State of Kerala Maradu Municipality & Ors. CIVIL APPEAL NOS.4784–4785 OF 2019 (Arising Out of SLP (C) Nos.4227–4228 of 2016).

Gaurika Chugh is a research scholar at the Centre for the Study of Law and Governance and Research Assistant at the Special Centre for Disaster Research, Jawaharlal Nehru University (JNU), New Delhi.

Farmers, Climate and Disaster Management in a Coastal Region

Swarnamayee Tripathy

Introduction

On a regular basis, we come across media reports of farmer distress across India. Nature, which is supposed to nurture them, has become the cause of their anguish and disappointment due to its wrath. Attempting to counter the fury of nature has been the reason for their suffering. In recent years, the rise in the number of farmer suicides in India has stunned the international community. Through their research, economists and sociologists have concluded that disasters, both man-made and natural, are one of the many important reasons behind this social phenomenon. Disasters result in crop loss leading to farmers' indebtedness, helplessness and consequently loss of lives.

India is one of the most disaster-prone nations on Earth. The unique sub-continental dimensions, geographical positions and the behavior of the monsoon make India one of the most hazard-prone countries in the world, with floods being the most common natural disaster in the country. The report, 'Decoding of Monsoon Floods', co-authored by Delhi-based

S. Tripathy (✉)
School of Social Sciences, Utkal University, Bhubaneswar, India

A. Singh et al. (eds.), *Development in Coastal Zones and Disaster Management*, Disaster Research and Management Series on the Global South, https://doi.org/10.1007/978-981-15-4294-7_7

NGO SEEDS and the Brussels-based Centre for Research on Epidemiology of Disasters,[1] 2018 finds that:

> Floods are amongst the most damaging and recurrent of all disasters; floods make up the highest number of disaster events in 2000-2017 across Bangladesh, India, Myanmar and Nepal and 55 percent of the natural disasters to strike India since 2000 have been floods. India had a mean of 11 flood events per district over the last 18 years. Ninety-eight percent of its 642 districts have received at least one flood event. Coastal floods accounted for only 1% of the flood events, while riverine foods accounted for the maximum. 71% of the floods are riverine floods in India. Even the hot deserts of Rajasthan known for drought have received more than the national average of 11 floods between 2000 and 2007. Floods are recurring across the region of South Asia. Flash floods that bring a level of unpredictability accounted for almost a sixth of the total. This has risked the lives and livelihood of the vulnerable most.

The farming community is the worst affected by these disasters, and aside from floods, drought conditions have also affected their livelihoods. Sixty-eight percent of cultivable land in India is vulnerable to drought.[2]River flooding in many areas deposits fertile sediments that aid in food production. But flash floods cause distress to the farming community. For states like Andhra Pradesh, Assam, Kerala and Odisha these disastrous events have taken a heavy toll on the states' economies as losses have been estimated in the millions of rupees. Hence, effective disaster management will to some extent save the country and the farming community from such agony.

Prior to the Yokohama conference (1994) disaster management was perceived as a short-term relief undertaking. Other pre- and post-disaster activities to mitigate the impact of disasters on specific communities were not construed as part of the larger framework of disaster management. The Yokohama Strategy and Plan of Action for a Safer World and the World Disasters Report (2002) raised the consciousness of the international community by highlighting the understanding that 'disasters pose a grave threat to the survival, dignity and livelihood of the individuals, particularly the poor and the hard-won developmental gains'.[3] Therefore, the Hyogo Framework for Action (2005–2015) adopted at the World Disaster Conference at Kobe, Hyogo, Japan emphasized 'Building the Resilience of the Nations and Communities to Disasters'. It defined 'disaster management' as a circular approach that includes both pre-disaster risk reduction and post-disaster administrative response. It recognized the need for good legislation. This was further strengthened by the Sendai Framework which identified four priority areas, namely, understanding disaster risk, strengthening disaster risk governance to manage disaster risk, investing in disaster risk reduction for resilience and enhancing disaster preparedness for effective response and to 'Build Back Better' in recovery, rehabilitation and reconstruction (Amita Singh et al. 2018). Other studies have found that documenting indigenous communities' coping mechanisms through research will add to the inventory of knowledge on disaster management. Therefore, in this discourse on disaster mitigation the more commonly used approach, 'disaster risk reduction'(DRR), is replaced by a more appropriate model, 'Resilience building, Risk reduction and Mitigation' (RRM).

India, in being pro-active, responded to the Hyogo Framework and institutionalized disaster management by enacting the Disaster Management Act, 2005. Over the past couple of years, the Government of India has brought about a paradigm shift in its approach to disaster management. The new approach is based on the assumption that development cannot be sustainable unless disaster mitigation is built into the development process. Another vital aspect of the approach is that mitigation has to be multi-disciplinary, spanning all sectors of development. The new policy also emanates from the belief that investments in mitigation are much more cost-effective than expenditure on relief and rehabilitation.[4]

In this context, this chapter attempts to examine the public policy on disaster mitigation with regard to farmers in coastal India.

Assumptions of the Study

Adaptation to climate change is a challenge for all countries. From a global perspective, the adaptation challenge is greatest for developing countries. They are more vulnerable to climate change because their economies are more dependent on climate-sensitive sectors, such as agriculture, fishing and tourism. With lower per capita incomes, weaker institutions and limited access to technology, developing countries have less adaptive capacity. Thus, it is assumed that pre-disaster administrative management in a disaster-prone region is intended to build in the requisite consciousness and confidence in the people and administration, as well as to manage mega-disasters.

Further, structure, density and socio-economic indicators of the population determine their level of vulnerability. Therefore, the nature of vulnerability has to be incorporated into public policy so that the fault lines of social justice are adequately addressed. In this context, it is pertinent to mention that the core capabilities of the vulnerable population need to be enhanced to meet disaster threats in a scientific way.

The Coastal Region of India

India has a long coast line of 7, 517 km. Nine states and four Union Territories come under the coastal region of India. The east coast lies between the Eastern Ghats ecosystem and the Bay of Bengal. The west coast strip extends from the Gulf of Cambay (Gulf of Khambhat) in the north to Cape Comorin (Kanyakumari). Five hundred and sixty million people live in this coastal region. One hundred and seventy-one million people in these coastal states are directly affected by floods and cyclones followed by torrential rain. In addition, approximately 4 million people comprising 8,64,550 fisherfolk households are affected by these disasters. Further, climate change related to sea level rise increases the vulnerability of the coastal eco-system by posing a threat to many coastal cities, urban centers and coastal populations in developing countries.[5] India is no exception to this.

Problems of Coastal Ecosystems

The changes observed along the coastline due to increasing use of coastal regions for developmental purposes such as industry, fishing and harbor facilities are as follows:

(a) Shifting of the shoreline;
(b) Coastal erosion;

(c) Formation of sand bars at inlets of the water body;
(d) Change in velocity and currents of coastal water;
(e) Littoral drift;
(f) Onshore/offshore sediment transport;
(g) Sand mining;
(h) Siltation;
(i) Change in beach profile.

Coastal Erosion

Coastal erosion is one of the main reasons for disaster in coastal regions. The coastal environment is a complex and dynamic ecosystem, in which there is a constant interaction between land, sea and air, culminating in shoreline changes. The landward displacement of the shoreline caused by the forces of waves and currents is termed 'coastal erosion' and is defined as the wearing away of the land. Although coastal erosion is a chronic problem, it is often considered a natural phenomenon inevitable along most open shores. While the effects of waves, currents, tides and wind are primary natural factors that influence the coast profile, various anthropogenic factors also trigger beach erosion. These include construction of artificial structures along the foreshore, mining of beach sand, offshore dredging and building dams across river basins. These developments are not compatible with the dynamic nature of the shoreline.

Erosion along Odisha Coast: A Case Study

Odisha has a coastline of 480 km, which is 8% of India's coastline. The vulnerability of coastal Odisha impacts 7168 villages, 20 towns, 10.60 lakh households and 58 lakh people within 25 km of the coastline.[6] Widespread erosion has been noted along approximately 187 km of the coastline. To cite a few examples, the continuous buffeting of the shore by the waves along Pentha village in Patkura Block and Satvaya village in Rajnagar Block of Kendrapada district of Odisha has resulted in the coastline receding landward by 200 to 300 meters.[7] This has been studied by the Geological Survey of India, revealing periods of accretion and erosion related in all probability to climate change, sea level fluctuations and human activity. Similarly, panic gripped the town of Puri, in 2016, when tidal waves washed away not only a part of the beach, but also the adjoining stretch of newly constructed road from Palm Beach Hotel to Sterling Resort, violating Coastal Regulation Zone norms. Moreover, major ecological habitats and delicate systems along the coast, such as the largest rookeries in the world of Olive Ridley sea turtles (the extensive sandy beach of Gahirmatha on Rushikulya); Asia's largest brackish water lagoon, the Chilika Lake; Mahanadi Delta, the second largest delta in India; and the extensive mangroves of Bhitarkanika, are also under constant threat from coastal erosion.

Therefore, climatic conditions caused by coastal erosion and the coping mechanisms adopted by people over the years need to be studied in detail while creating policies for disaster mitigation.

Farmers in India

Farming is the main production activity of the majority of the rural population of India. People working in the rural ecosystem are dependent on farming for their livelihoods as farmers or farm-laborers, and their well-being is closely related to farm production. As per the Registrar General of India & Census report 2011, the total farmer or cultivator population of India is 118.7 million (2011 Census) and that of farm laborers/agricultural laborers is 144.3 million, comprising 31.55% of the total rural population. The Census of India 2011 reports that cultivators constitute 24.6% of the 481 million-strong workforce of India. Approximately 57.8% of India's rural households are engaged in agriculture. However, when disasters strike a particular region, the livelihood of both cultivators and agricultural laborers is affected, pushing them further into the trap of misery and poverty.

Under the Agricultural Debt Waiver and Debt Relief Scheme 2008, farmers in India are categorized into three types depending on their operational landholdings. Operational holding refers to all land which is used wholly or partly for agricultural production and is operated as one technical unit by one person alone or with others without regard to the title, farm size or location. 'Marginal farmer' is the first category,

Table 1 Farmers in India, 2016

Type of farmers	*Size of land-holdings (in hectare)*	*% of farmers (2010–11)*	*% of farmers (2015–16)*
a) Marginal farmers b) Small farmers	1 h. 1–2 h.	86.21	84.97
a) Semi-medium farmers b) Medium farmers	2–4 h. 4–10 h.	13.22	14.29
Large farmers	More than 10 h.	0.57	0.71

Source: Agriculture Census of India, 2015–16

cultivating (as owner or tenant or sharecropper) agricultural land up to 1 hectare (2.5 acres). 'Small farmer' refers to a farmer cultivating (as owner or tenant or sharecropper) agricultural land of more than 1 hectare and up to 2 hectares (5 acres). 'Other Farmer' refers to a farmer cultivating (as owner or tenant or sharecropper) agricultural land of more than 2 hectares (more than 5 acres). As much as 67% of India's farmland is held by marginal farmers with holdings below 1 hectare, against less than 1% in large holdings of 10 hectares and above, the latest agriculture census (2015–16) shows. The percentage of female operational landholders increased from 12.79% in 2010–11 to 13.87% in 2015–16. Table 1 presents a picture of the nature of the farming community in India.

According to the Agriculture Census (2015–16), the total number of operational holdings in India numbered 138.35 million with an average size of 1.15 hectares. Of the total holdings, 85% are in marginal and small farm categories with less than 2 hectares of land and are engaged exclusively in livelihoods such as livestock, poultry and fishing. The estimates indicate that small and marginal farmers may account for more than 91% of farm holdings by 2030, and the average size of these holdings has shown a steady declining trend over various Agriculture Censuses since 1970–71.

Another cause for concern is that in 2010–11, the proportion of net irrigated area to net area sown was 45.70%, which shows that half the country's farmland relies entirely on rains for their crops. To cite the example of Odisha, with a 480 km coastline that is prone to climate-mediated cyclones and coastal erosion and with water resources dependent on monsoons, agriculture in the state is more vulnerable to climate change. Water-consuming rice is Odisha's main crop and therefore its farming community is vulnerable to the vagaries of climate-induced weather changes.

Impact of Disasters on the Farming Community in India

All disasters have social, economic and health consequences of varying magnitude. They affect men, women and children at different magnitudes depending on their level of resilience to counter social and economic losses. This is directly related to people's economic status and nature of their livelihood. Farmers in coastal regions experience frequent flash floods throughout the year which result in heavy loss of land and property and, in turn, push them into a vulnerable state and at times compel them to end their lives. It is pertinent to mention here that in 2014, the National Crime Records Bureau of India reported 5650 farmer suicides.[8] The highest number of farmer suicides was recorded in 2004 when 18,241 farmers committed suicide. The farmer suicide rate in India ranged between 1.4 and 1.8 per 100,000 total population, over a 10-year period through 2005.[9] It can be presumed that hardships caused by disasters are one of the causes of crop loss for farmers, pushing them into hardship which ultimately leads to suicide.

Agriculture plays a vital role in the Indian economy. Over 70% of rural households depend on agriculture as their principal means of livelihood. Agriculture, along with fisheries and forestry, accounts for one-third of the nation's GDP and is its single largest contributor. Thus, while considering the impact of disasters we need to take into account the farmers as well as the agricultural laborers who depend on agriculture for their livelihoods. As per the census of 2011, 263 million people are engaged in the agriculture sector and over half of them are agricultural laborers. If a disaster occurs and impacts agriculture, the livelihoods of all these people are affected, pushing them into the trap of poverty.

A farm laborer who works for daily wages must regularly look for work. When disaster strikes the area and agricultural activities come to a halt, farm laborers choose to migrate for want of livelihood and are exploited as migrant workers. In addition, the work opportunities for farm laborers have reduced due to mechanized farming. Marginal and small farmers do not engage farm laborers in their fields as they themselves do all types of farming activities. Farm laborers are usually employed by medium and large farmers.

Dairy farming is a common farming activity for many rural families and supplements the family income. Disasters can affect this small part of a family's income. A disaster also affects the small household manufacturing units which once added to the family income, such as papad-making and beedi-making.

The farming community and the people attached to agriculture for their livelihood develop a cultural attachment to the river in their ecosystem. They continue to live in these areas because of their deep attachment to the river ecosystem, despite continuous adversities posed by the river. Instead, the inhabitants develop coping mechanisms for living in floodplain areas with the help of traditional knowledge systems. This poses a challenge for policy-makers in terms of considering the sentiments of stakeholders and at the same time devising effective policies and mechanisms to cope with disasters.

Public Policy for Farmers in Disaster-Affected Regions and the Policy Gaps

Ironically, countries which are disaster prone do not mention of disaster management as a subject in any of the three lists referred to in the 7th Schedule of the Constitution of India. A subject that is not specifically mentioned in any list has to be taken care of by the Union Government as per Entry 97 of the Union List. But the basic responsibility for undertaking rescue, relief and rehabilitation measures in the event of natural disasters belongs to the state government concerned. The role of the Central government is supportive, in terms of physical and financial resources. However, 29 items in the 11th Schedule and 18 items in the 12th Schedule have gone a long way to ensuring better disaster mitigation.

Disasters have long posed a threat to society and the economy, and therefore disaster management has occupied a pivotal place in development planning. Disaster management is no longer a piecemeal strategy and has become an integrated process and approach since the Hyogo Framework for Action, 'Building the Resilience of Nations and Communities to Disasters (2005–2015)'. Following this, India issued the Disaster Management Act 2005 and related legislation in its federal units. Peoples' perspective were incorporated into this policy after the Sendai Framework for Disaster Risk Reduction (2015–2030), which has as its top priority 'strengthening disaster risk governance' and 'legal frameworks' (Amita Singh 2018, p. 1).

The Disaster Management Act, 2005, of India clearly specifies the role of states in putting in place an inclusive policy for disaster management. Section 23(4) of this Act emphatically specifies that the State Plan shall include:

- The vulnerability of different parts of the state to different forms of disaster;
- The measures to be adopted for the prevention and mitigation of disasters;
- The manner in which the mitigation measures shall be integrated with development plans and projects;
- The capacity-building and preparedness measures to be taken.

Accordingly, the Government of India has taken some administrative steps to reduce the miseries of farmers in a post-disaster situation. One of these is insuring crops. Many parts of India experience unseasonal dust and thunderstorms, followed by unseasonal rains year after year. This has cost lives and led to extensive crop damage. With freak weather events becoming more common, protection of farmers through crop insurance has become an essential component of national agricultural policy since the 1960s (Prabhu and Ramachandran 1986; Sinha 2004; Vyas and Singh 2006; Raju and Chand 2007; Nair 2010). From the Comprehensive Crop Insurance Scheme (1985) through to the National Agriculture Insurance Scheme (1999–2000), the Weather Based Crop Insurance Scheme (2003), the Modified National Agriculture Scheme (2010) and through to the

Pradhan Mantri Fasal Bima Yojana (PMFBY) (2016), India's agricultural insurance schemes have undergone several changes in their approaches. Mitigating risk in the farm sector has direct implications for agricultural productivity and farmers' well-being. However, only 10% of farmers have been covered under this since the Pradhan Mantri Fasal Bima Yojana was introduced. Previously it was only 2% and covered risk assessment of individual farmers as per the areas cultivated, consequently benefiting the medium and large farmers more. Economists like Dandekar have suggested adopting an 'area approach' for risk assessment to ensure equity through the scheme.[10]

The PMFBY scheme is a multi-peril crop insurance scheme introduced in January 2016. It follows an 'area basis approach' with seasonal activities under the consideration of a broad set of risks spanning various stages of crop development and post-harvest losses due to natural calamities. States and union territories have been entrusted by the State-Level Coordination Committee on Insurance to notify the insurance unit such as village/village panchayat or any other equivalent unit for major crops. The maximum premium payable by the farmers is 2% for all kharif food and oilseed crops, 1.5% for rabi crops, and 5% for annual commercial and horticultural crops. It aims to reach 50% of farmers in India.

The most important aspect of this agricultural insurance policy is that it is mandatory for those farmers who require agricultural credit from banks. Other farmers may voluntarily opt for this credit by paying a premium. The low percentage of farmers in the insurance net reflects the low penetration of index-indemnity insurance product.[11]

Further, state governments have their own policies within the broad framework of the administrative design of the Union. Notwithstanding the marked progress following the adoption of the area-based approach in crop insurance schemes, the credit-linked insurance scheme design and distribution, reliance on informal credit, and exclusion of non-loanee smallholder farmers have restricted the scope for broad-based, yet voluntary, participation of farmers and other stakeholders.

Disbursal of compensation for crop loss of farmers is carried out based on ownership of land. A farmer holding Record of Right (ROR) receives compensation for crop loss whereas the tenant or sharecropper who actually cultivates the land gets nothing. With regard to Odisha, delivery of crop insurance has been linked to ROR on farmers' land. Thus, tenants or sharecroppers who cultivate farmland of urbanite large and medium farmers do not receive crop insurance. Rather, the insurance comes to the bank account of the owner of the land. By virtue of a letter issued by the Special Relief Commissioner, Disaster Management, Government of Odisha, on August 18, 2015,[12] a sharecropper is entitled to receive crop-loss compensation based on the on-the-spot eye-estimation report given by the Revenue Inspector of the respective area. This government circular has opened the door for corruption by the Revenue Inspector/Amin rather than helping the tenant or sharecropper in a real sense. This policy gap has added to the miseries of sharecroppers in post-disaster situations. Though more than 85% of the land is cultivated by sharecroppers in Odisha, they do not receive any government assistance at the time of natural calamities, nor do they receive loans from banks or fertilizers at subsidized rates, because there is no such provision under the existing law. The sharecroppers are also not entitled to obtain input subsidies, insurance coverage, relief or other benefits. The Task Force on Agricultural Development set up by the NITI Aayog identified the absence of formal recognition of tenancy in many states as a major hurdle in consolidating landholding.

The disaster mitigation approach of India has exclusionary trajectories as successive policies have paid very little attention to the woes of disaster-affected farm laborers. The laborer who lives in a kutcha house with subsistence-level livelihood goes without work day after day in a post-disaster situation. After a period of coping with bad times, the laborer opts to migrate to an unknown destination, leaving his family members behind. Research findings show their exploitation after migrating.

Developmental Challenges to Public Policy-Making

Socio-economic vulnerabilities caused in a post-disaster situation are not similar across all social classes. If the proper policy is not put in place and specific vulnerabilities of different areas and people are not addressed, the concern of human rights is also not properly addressed. Therefore, this study puts forth the following suggestions:

- Development of flood-resilient agriculture is one of the possible solutions. Research on short-duration crops for disaster-prone zones of India has become essential.
- Farmers in most parts of India adopt mono-cropping as they are at the mercy of monsoon. Developing the capabilities of farmers by providing inputs to opt for multiple-cropping and cultivating short-duration crops is a practical solution for risk reduction.
- In spite of the government's good intentions, assessments of the PMFBY face several challenges that make processing and verification of insurance claims error-prone and time-consuming. Payouts do not reach farmers at the right time and in amounts commensurate with their losses. Many experts and organizations working in this area are now recommending the use of information and communication (ICT) tools to help farmers regain faith in crop insurance schemes and make them more efficient and transparent.
- Policies of crop insurance have to be redefined to include sharecroppers. The present government of Odisha has addressed this issue in an ad hoc way. Official sources state that 55 lakh families in Odisha are farmers, and of these 80% are sharecroppers, the real tillers of the land. In the state, produce is distributed between sharecroppers and landowners. Crop-loss compensation was given to landowners starting in 2015 and the practice continues to date. As per the letter issued by the Relief Commissioner in 2015, crop-loss compensation is calculated based on eye estimation or crop-loss certification by assessing crop loss by crop cutting after any disaster. However, very few sharecroppers have received this compensation, and those who have received it have done so by bribing the revenue inspector, says one land owner of Tholanda Grampanchayat of Jagannath Prasad Block in Ganjam district of Odisha. The same opinion has been expressed by sharecroppers. One of the landowners in the above-mentioned panchayat expressed that they are ready to enter into an agreement with the sharecroppers on a yearly basis. But their fear is that the government would be able to alter the data on the computer on a yearly basis, and the landowners have no faith in the administration. Therefore, Odisha has failed to bring about a land lease legislation due to lack of consensus. Effective communication is essential to give the message to the landowners that their land is

protected and cannot be taken away by the cultivators. This will be the biggest incentive for them to reveal the identity of the sharecroppers who are working over the farm but do not own the farm.

- The Odisha government, however, is planning to empower the sharecroppers without disturbing the existing land ownership pattern. The terms of agreement should be for a minimum period of three years, and the parties with mutual consent may further extend it. The lessee cultivator should not have any right over the leased agricultural land except for cultivation purposes. The landowner will be free to mortgage the land without affecting the agreement and will not be under any obligation to extend the contract. In the event of any dispute between the cultivator and the landowner, they will resort to the local dispute resolution mechanism and will not approach the judiciary. Local revenue officials, who will register the deeds, will play a major role. Once the tenants are identified and registered, the government will have no problem in extending financial assistance to the cultivators. The government should stress effective communication. The respective landowner will thus have distributed the compensation amount among the sharecroppers in proportion to the amount of land cultivated by the sharecroppers.
- The government should encourage farmers to opt to insure their crops. Awareness among farmers will take them a long way in combatting the difficulties of crop loss.
- The farmers in disaster-prone areas should be given a comprehensive farm kit in a post-disaster period which would include seeds, fertilizer and other farm necessities required for short-duration crops to ride out the difficult situation.
- What is required is a prior mapping of the productive potential of the disaster-prone area and the capabilities of its youth and adult population. Accordingly, strategies for skill development for youth and adults have to be introduced in such areas prior to the onset of disasters. As an example, in a post-disaster situation, a young farmer can buy sugarcane, make jaggery and sell it in the nearby urban market.

- Banks should be directed to offer small loans without any collateral security to landless youth to buy vehicles for transportation. This will supplement his household income in a post-disaster situation.

Conclusion

The post-Sendai Framework has established the interconnection between pre-disaster risk reduction through building community resilience and post-disaster rehabilitation in a larger framework of disaster governance. It emphatically states that better preparedness ensures quicker response, which means less disaster loss, better mitigation and preventive activities, and better protection against hazards. The concept of the Disaster Management Cycle is pertinent to public policy if people's perspective is given due attention while drafting the policy. Past disaster management has either not taken into account any policy matters or not accorded them sufficient primacy especially in poor developing economies where there are other pressing concerns to address such as poverty, malnutrition and unemployment, and other development issues. However, losses suffered around the world due to natural disasters have brought about the realization that development cannot be sustained without effective disaster mitigation efforts. Since 'a stitch in time saves nine', this needs to be driven home to policy-makers to factor disaster management concerns into development planning.

Notes

1. 'Decoding the Monsoon floods in Bangladesh, India, Myanmar and Nepal (2017)', Report prepared by SEEDS, New Delhi, and Brussels-based Centre for Research on Epidemiology of Disasters (CRED).
2. National Policy on Disaster Management (October 22, 2009), retrieved from http://ndmindia.nic.in/NPDM-101209.pdf . p.1.
3. Singh, Amita, (2018), *Disaster laws: Emerging Thresholds*, Routledge.
4. Govt. of India, (2004), 'Disaster Management in India: A Status Report', Ministry of Home Affairs, National Disaster Management Division, pg. 4.

5. Senapati S. and Vijaya Gupta, (2014) 'Climate Change and Coastal Ecosystem in India: Issues in perspectives', *International Journal of Environmental Sciences*, Vol. 5, No 3. The authors argue that climate change affects the coastal ecosystem and the livelihood of the fishing community.
6. Kumar, T.S. et al. (2010), 'Coastal Vulnerability Assessment for Orissa State, East Coast of India', *Journal of Coastal Research*, 26(3), 523–534, Florida.
7. Senapati Ashis, (Oct.2018) 'Climate change in India: Odisha all set to redraw its map', https://www.downtoearth.org.in/news/climate-change/climate-change-in-india-odisha-all-set-to-redraw-its-map-61873
8. National Crime Reports Bureau, 'ADSI Report Annual – 2014'. Government of India, p. 242, Table 2.11
9. Gruère, G. & Sengupta, D. (2011), 'Bt cotton and farmer suicides in India: an evidence-based assessment', *The Journal of Development Studies*, 47(2), pp. 316–337.
10. http://timesofindia.indiatimes.com/articleshow/66035943.cms?utm_source=contentofinterest&utm_medium=text&utm_campaign=cppst
11. *Crop insurance scheme has been defined as index-indemnity insurance product by economists.* Indemnity insurance is a contractual agreement in which one party guarantees compensation for actual or potential losses or damages sustained by another party. These special insurance policies indemnify or reimburse professionals against claims made as they conduct their business. *Index linking* is used in a variety of property *insurance* policies, including home buildings and content insurance as a way of ensuring that the sums *insured* under the policy are kept in line with current market day values.
12. Govt. of Odisha letter no.3654 R & DM(SR) dated August 18, 2015, issued by Special Relief Commissioner, Disaster Management.

References

Nair, R. (2010). Crop insurance in India: Changes and Challenges. *Economic and Political Weekly, 45*(6).

Prabhu, K.S., & Ramachandran, S. (1986). Crop Credit Insurance: Some Disturbing Features. *Economic and Political Weekly. XXI*(42).

Raju, S.S., & Chand, R. (2007). Progress and Problems in Agricultural Insurance. *Economic and Political Weekly, 42*(21), 1905–1908. Jan. https://doi.org/10.2307/4419629

Sinha, S. (2004). Agriculture Insurance in India: Scope for Participation of Private Insurers. *Economic and political weekly 39*(25), 2605–2612. Jan. https://doi.org/10.2307/4415175

Singh, A., Punia, M., et al. (2018). *Development and Disaster Management*. London: Palgrave Macmillan.
Vyas, V.S., & Singh, S. (2006, November). Crop Insurance in India- Scope for Improvement. *Economic and Political Weekly, 41*.

Swarnamayee Tripathy is a professor and head at the Department of Public Administration, Utkal University, Bhubaneswar, Orissa.

PART III

Conserving Marine Flora and Fauna

Marine Animals and Coastal Disasters

Amita Singh

Introduction

A healthy habitat has a wholesome animal population which ensures sustainability and a stable environment relatively free of many natural disasters. Earth is a protective blanket of interconnected ecosystems which provide sustainability to human progress. These ecosystems are intertwined and survive interdependently. Ironically, life is an unfolding story of these ecosystems yet misguided development has been consistently destroying them and creating increased vulnerability to disasters. According to the UNEP, sustainable economy is worth 12 trillion USD and has the potential to create 380 million jobs. In 2020, when world leaders meet in China to strike a new deal for conserving nature, only a quarter of land on earth is substantially free of human activities. This is further projected to decline by one tenth by 2050, rendering the planet defenseless against destruction and disasters. To reverse the countdown, UNEP's mandate puts forth 'five transformations' to recalibrate humanity's relationship with nature and harness nature-based solutions for climate change. In all five transformations some attention to animal ecosystems is unavoidable (Fig. 1).

A. Singh (✉)
Centre for the Study of Law & Governance, Jawaharlal Nehru University, New Delhi, India

A. Singh et al. (eds.), *Development in Coastal Zones and Disaster Management*, Disaster Research and Management Series on the Global South, https://doi.org/10.1007/978-981-15-4294-7_8

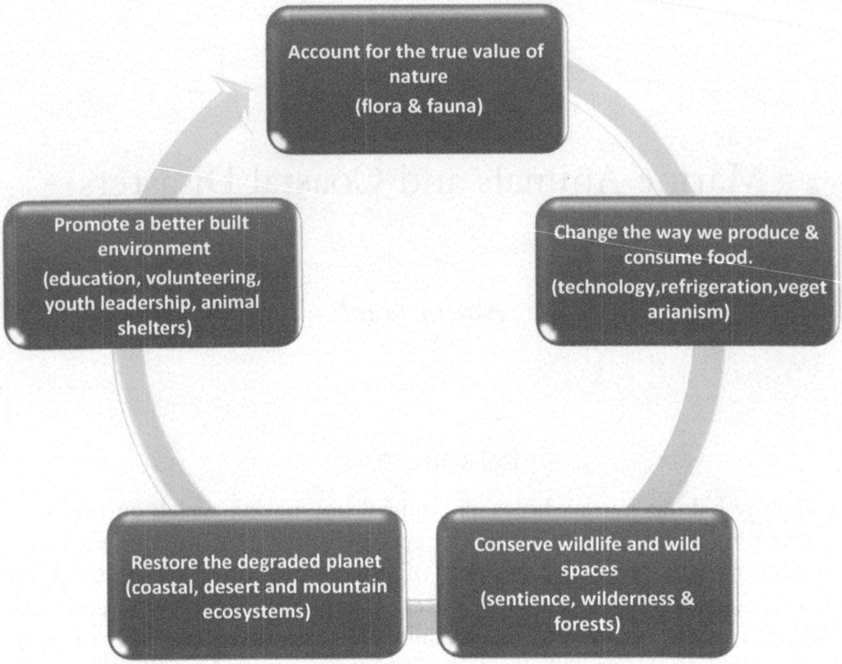

Fig. 1 Five transformations to relate to nature (bracketed content is interpretative of the UNEP's suggested transformation). (Source: UNEP https://wedocs.unep.org/bitstream/handle/)

Why Address Animals in Disaster Studies?

Disaster studies is undergoing a major transformation with an improved understanding of the nonhuman environment. Sociological and ethnographic research, the spread of media and a progressive advance of sentience as objective analytical tools and techniques of social behavior challenge a formerly held pre-determined and socially constructed view of environment and vulnerability. Disasters today are not simply an unintended calamity provoked by natural consequences and responded to through the demand for an efficient state action. Human beings are recognized as having intricate linkages within a web of life in nature, which they conserve and also exploit, generating hazards for the rest of the ecosystem. The emergence of Environmental Sociology highlights biotic relationships which sustain human life but through ignorance, greed or wrong decisions turn a hazard into some of the worst disasters. While liberating human sentience from the cudgels of deterministic policies, Catton and Dunlap (1978, p. 45) identified how the limits of ecological resources strain human relationships within the 'intricate web of cause and effect and feedback in the web of nature produce many unintended consequences from purposive human action'. There is a wide literature which spells out the logic and need for approaching disasters with better sociological tools and incorporating human and nonhuman connections in sociological research. An increasing number of sociologists are emphasizing, through the American Sociological Association's section on Animals and Society, the Australian Sociological Association and the International Sociological Association, recognizing the importance of animals to societies and the rights of nonhumans to be treated at par with humans (York and Longo 2017; DeMello 2012; Peggs 2012; York and Mancus 2013; Surrallés 2017). Most interestingly, many sociologists have broken the human exceptionalist paradigm by bringing in 'spirituality', as in *Surrallés*'s 'Human rights for nonhumans?' and changing the notion and nature of international law following the 2007 United Nations Declaration on the Rights of Indigenous Peoples.

A focus on anything beyond a human-centric paradigm raises many epistemological questions and it is challenging to answer them through scientific logic, especially within the limited period of a natural catastrophe. Many of these answers would be new to the established conventional disciplines of disaster studies and disaster law but would also develop a new sociological methodology in which it becomes absurd to even think about human existence in a society without its relationship to all nonhumans sharing a habitat with humans. This demands a holistic vision of extreme democratic enlightenment where the boundaries of assessing vulnerability transcend the boundaries of not just socially vulnerable groups such as race, gender, caste and religion but also speciesism.

Do Global Environmental Indicators Address Nonhuman Concerns?

An environmental indicator is a metric describing the health of the ecosystems in a particular area. This suggests the minimum capacity to allow other dependent and interconnected communities to perform their ecosystem services.

A core of global environmental indicators from 1990 to 2005, produced by UNEP, shows the insensitivity of a global environmental regime to a major component of environmental health. The total benefit of carbon sequestration and coastline management as contributed by coastal and marine animals finds no replacement in human technology or scientific design. Greenhouse gas and CFC emissions increase phenomenally with the loss of coastal animal life. However, the eight indicators, that is, energy, atmosphere, biodiversity, ozone depletion, forests, freshwater, oceans and governance, are so fuzzy that they leave ample scope for the policy-maker to display a failed policy in an area as a success. Energy supply isolated from granular data on thermal, hydro and non-conventional resources may not be enough to explain the environmental contribution of this indicator. Similarly, the atmosphere is not simply gases, as much as biodiversity is not species alone. The interlinkages among ecosystems, as well as extinction and revival of species, explain how a particular variable can be an indicator of the global environment. This chapter will explain the interlinkages of coastal ecosystems in which animals are one of the most powerful factors in environmental health and disaster proofing along the coasts.

The UNEP's 'Core Set of 8 Global Environmental Indicators' are primarily anthropocentric, that is, whatever humans consume is to be prioritized for conservation, such as *Energy*, *Atmosphere*, *Biodiversity*, *Ozone Depletion*, *Forests*, *Freshwater*, *Oceans and Governance* (http://www.grid.unep.ch/). A concern for coastal wildlife and sea animals would have made the narrative more inclusive. Once indicators are set at the top UN body, all countries and their local administrative agencies copy the instructions and carry them further. Including animals as one of the major indicators for disaster in the making is imperative for a more evolved science on disasters.

Aggrandizing Humans and Eroding Coastline

It is injudicious to estimate the role and contribution of coastal animals to coastal ecology without referring to their habitats. The habitats of marine animals along a damaged coastline are nearing a terminal stage. There are many anthropogenic causes which have led to habitat destruction, and these are not debatable, not even the sea level rise which the coastline states have to accept and adjust to as reality. Global Mean Sea Level has risen by 10 to 20 cm in the past 100 years. The annual rate of rise over the past 20 years has been 3.2 mm, which is twice the average rate of the preceding 80 years. The trend could threaten habitations along India's coastline. This would increase flooding from 300 to almost 900% and will severely affect the approximately 600 million people living in the nine coastal states, two Union Territories and two island states in India. Out of the 10 most vulnerable cities in the world, India's Mumbai and Kolkata occupy the second and third ranks after China's Guangzhou (Singh 2018).

Adding to the coastline damage due to sea level rise is the rising density of populations in coastal regions of the world. India's coastal population has exploded to approximately 600 m in the last decade alone. The 8000 kms of coastline full of wetlands, lagoons, mangroves, seagrass beds, coral reefs, bays, creeks and estuaries has deteriorated into a broken, segmented, washed off, starving and drought-hit rim due to indiscriminate growth, construction and forest destruction. The worst aspect of this in terms of coastal habitat is the loss of a unique home for millions of animals, the mangrove forest. Mangroves are a type of halophytic vegetation that grow in the tropical and subtropical regions in the latitude 24°N—38°S. They have almost floating succulent leaves, sunken stomata and stout and strong aerial breathing roots. They can withstand strong waves, tsunamis and erosion and thus protect the many smaller marine creatures within their cozy root caves shaped artistically with pneumatophores. They have a high salt concentration and support structures such as stilt roots and buttresses. Since mangroves are viviparous, and germinate within the fruits which are still attached to the parent plant, this ensures continuity and a well-knit complex root system which facilitates accumulation of organic detritus[1] and inorganic nutrients, thus acting as a breeding and nursing ground for various marine and pelagic fishing. They also act as a zone of land accretion when, due to the gradual and imperceptible natural deposition and trappings of fine sediment entangled in mangrove roots, more land is added to the coastline. This accretion acts as a shield against sea level rise in addition to protecting against storm surges and tsunamis. After the 2004 tsunami disaster-related loss of mangroves has become a critical concern.

Mangroves are a world of animals, a source of livelihood for millions, food security for coastal populations and a wall against disasters. The National Remote Sensing Agency sent a warning signal when it recorded a decline of 59.18sq.km of mangroves between 1972 and 1975 and 1980 and 1982.[2] India has lost 40% of its mangroves in the last century.[3]The hurricane-ridden east coast of India has lost 26% while the flood-vulnerable

Table 1 Total mangrove cover since 1987

s.no.	State/UT	State of forest report Total mangrove cover			
		1987	1989	1991	1993
1	Andhra Pradesh	495	405	399	378
2	Goa	0	3	3	3
3	Gujarat	427	412	397	419
4	Karnataka	0	0	0	0
5	Kerala	0	0	0	0
6	Maharashtra	140	114	113	155
7	Odisha	199	192	195	195
8	Tamil Nadu	23	47	47	21
9	West Bengal	2076	2109	2119	2119
10	A&N	686	973	971	966
11	Daman & Diu	0	0	0	0
12	Puducherry	0	0	0	0
	Total	**4046**	**4255**	**4244**	**4256**

Source: Forest Survey of India, http://fsi.nic.in/isfr2017/isfr-mangrove-cover-2017.pdf

west coast has lost 44%, while the fragile coastal habitat of the Andaman and Nicobar Islands has lost more than 32% of mangroves.[4] Tables 1 and 2 present the brutal destruction of mangroves in 8 out of a total of 12 mangrove areas (9 states, 2 island states and 1 union territory) have zero dense mangrove cover, moderately dense area is also a minuscule of 1480sq.km, and open mangrove area which generally is used for cosmetic purposes of tourism and private cultivation is 1960 sq.km. Therefore the government claims that the mangrove area has increased from 4256 sq.km to 4921 sq.km, showing a change of almost 181 sq.km. This renders no habitat for animals, birds and fish. Concomitantly, it fails to provide other ecosystems and social services to coastal populations.

Mangrove deforestation has led to the destruction of coral reefs around its major formation areas in Lakshadweep, the Andaman and Nicobar Islands, the Gulf of Mannar and the Gulf of Kachchh. Due to intensive mining, dredging, fishing, coral collection and tourism large quantities of mud have been deposited over the reef, making it infertile and defenseless. With the death of mangroves and coral, the food of marine predators and large fish is lost and this is gradually leading to their reduced numbers and perhaps to their extinction. A survey of the Alappuzha coastal fisherfolk markets revealed that the top 20 big fish which provide them with a

Table 2 Total mangrove cover 2017

s.no.	*State/UT*	*Very dense mangrove*	*Moderately dense mangrove*	*Open mangrove*	*Total*	*Change with respect to ISFR, 2015*
1	Andhra Pradesh	0	213	191	404	37
2	Goa	0	20	6	26	0
3	Gujarat	0	172	968	1140	33
4	Karnataka	0	2	8	10	7
5	Kerala	0	5	4	9	0
6	Maharashtra	0	88	216	304	82
7	Odisha	82	94	67	243	12
8	Tamil Nadu	1	25	23	49	2
9	West Bengal	999	692	423	2114	8
10	A&N	399	169	49	617	0
11	Daman & Diu	0	0	3	3	0
12	Puducherry	0	0	2	2	0
	Total	**1481**	**1480**	**1960**	**4921**	**181**

Source: Forest Survey of India, http://fsi.nic.in/isfr2017/isfr-mangrove-cover-2017.pdf

healthy cash return are no longer seen even though these fisherfolk regularly fish along the seashore. Kerala has now lost all its mangroves. Some leftover patches of this water shrub are controlled by the hoteliers and private lodges and are no longer available to fisherfolk. The story of fisherfolk was not much different at Sundarbans as they lost their major livelihood to honeycombs and firewood.

Marine Animals and Carbon Sequestration Services

Studies suggest that coastal predators[5] such as whales, sharks, turtles, alligators, seals and walruses protect carbon sequestration in plant biomass and also in coastal soils. As pollution, land erosion and climate change increase, this sequestration process is weakened, resulting in coastal cascades, disasters and also hurricanes.

Coastal vegetation such as mangroves, salt marshes and seagrasses 'capture and hold' carbon underground as carbon sinks. This carbon is also referred to as 'Blue Carbon'. Despite its direct contribution to disaster prevention by locking up the release of CFCs, the coastal Blue Carbon ecosystems are the world's most threatened. Over 67% of mangroves, 35% of tidal marshes and 29% of seagrass meadows on average are brutalized, wiping out 340,000 to 980,000 hectares of Blue Carbon annually.[6]With this rate of destruction it is impossible to prevent a major fatal coastal disaster before the end of this century.

Blue Carbon is primarily created by sea animals. They sequester carbon below the ground and much faster than other ecosystems and for a much longer time, stretching into many millions of years. This carbon is found in coastal soil and is unleashed as free carbon which goes back into the environment and is emitted as greenhouse gases into the atmosphere. Carbon sequestration is performed only when these coastal animals are able to live in harmony with coastal vegetation. This prevents climate change and also ensures a healthy industrial policy by using these areas as carbon credits in carbon markets. By offsetting effluent discharge and pollution and promoting conservation of coastal animal and forest life, many direct or indirect pollution taxes are reduced and an incentive is created to restore and conserve animals, mangroves and other coastal vegetation. Yet, coastal developmental plans fail to acknowledge that loss of these animals disrupts carbon sequestration and subsequent loss of vegetation over a rich carbon coastal soil.

Studies on coastal ecosystems by leading ecologists have found that an estimated habitat loss of around 50% has already taken place (Zedler and Kercher 2005), but coastal predators, marine mammals and elasmobranchs patrolling the waters and sea birds over it have declined by 90–100% (Atwood and Hammill 2018, p. 2). Further, ecologists have presented the revealing finding that as predator declines habitat declines. This not only suggests but highlights for the information of all Coastal Disaster Management Authorities that '*predator declines is one of the surest symptoms of an impending ecological crisis at the coast which would sooner or later be triggering a disaster of any magnitude*'.[7] The 2018 floods in Kerala were predicted in studies conducted in 2015 on a ground-based study of its land management and ecosystem stress due to loss of ocean fish, diminishing coastal forests, unbridled construction, arbitrary tourism and failure to observe coastal zone regulations to protect fragile coastal plant and animal ecosystems (Singh 2016).

Distress and Killings of Coastal Animals

Unregulated mass killings of many threatened species in Indian coastal areas is posing a danger to seagrasses and a rapid advance of sea level rise due to climate change. The dugong an endangered herbivorous mammal which is listed as 'vulnerable' in the IUCN list with just 300 individuals left in a shrunken habitat of Gulf of Kachchh in Gujarat and the Palk Bay between India and Sri Lanka. Human habitation is increasing and the animal is being killed barbarically for meat. South Asian environmental efforts to protect this animal are weak despite campaigns and people's activism to save them. Such actions speed up climate change impacts in the region is a herbivorous mammal found only in the Gulf of Kachchh in Gujarat and the Palk Bay that separates Sri Lanka from India, and are being killed barbarically for meat but also threatening to speedup climate change and impending disasters.

Sea animals have been victims of many catastrophic and repeated global disasters such as oil spills. The oil spill resulting from the BP *Deepwater Horizon* catastrophe in Florida in 2010 and the oil spill in the Sundarbans' fragile and endangered animal zone in India (2014) have raised some crucial questions. Offshore drilling, transportation and storage have caused coastal animal and bird life to be disturbed. The BP *Deepwater Horizon* catastrophe killed 82,000 birds of 102 species, 6165 sea turtles and more than 25,900 marine mammals. The oil spill, affecting more than one thousand miles of coastline, destroyed beyond replenishment the total coastal ecosystems. In the Sundarbans, the oil spill caused by the Bangladesh Petroleum Corporation killed dolphins, otters, crocodiles and crabs and pushed to grave risk several species of endangered animals listed in the International Union for Conservation of Nature endangered animals list. Many bird species, such as the black-capped kingfisher and great egret, were almost wiped out. It was also found that the secondary effect of this oil spill was even worse in terms of species destruction and disease (Rahman and Rakhimov 2015).

Many coastal species are destroyed or displaced due to unusual and unprecedented ecological calamities. After the 2011 Japanese tsunami, a strange phenomenon was discovered regarding coastal animals. Hundreds of species from the Japanese coastal waters were carried across the Pacific on floating debris from the destruction caused by the tsunami. North American coastal life was heavily transgressed due to the invasion of a large number of mussels, sea slugs, sea anemones and crabs from across the Pacific. This little researched ecological phenomenon was not just a fallout of natural disaster but was primarily due to human behavior. This large-scale species migration from one habitat to unknown new destinations was possible due to plastic sheets and fiberglass which were durable flotillas that could carry entire families of sea animals without being destroyed. Professor Steven L. Chown at Monash University, Australia called it a 'process of mega rafting' (Fackler 2017).

Human insensitivity and ignorance about animals has also caused extreme brutality to animals trapped during disasters. During the raging floods at Levee in southeastern Iowa more than 36,000 pigs were washed away. Approximately 16 pigs swam to safety over a rooftop but were shot dead by the Des Moines County Sheriff's officials because their hoofs would have damaged the plastic roof (*The Spokesman Review* 2019). There are innumerable such cases from other disasters in which intransigence of decision makers becomes a cause for a bigger and equally damaging moral disaster embedded in hateful speciesism.

There are some good practices which also require the attention of policy-makers so that upscaling and efficiency feeding can improve the projects. Many lawmakers and legislators have acknowledged and supported disaster-trapped ocean animals. Some of these smaller victories are mentioned below to give an idea of the future role that humankind can play in pledging commitment to coastal protection:

1. Anti-fracking: A new scientific shale rock oil drilling technique from the deepest parts of the ocean is called 'offshore fracking'. Fracking refers to 'hydraulic fracturing' of rocks. Drilling companies with government support in the UK and USA are insisting that shale gas is recoverable from trillions of cubic feet below the ocean floor where regular human technologies may not work. The rock is pressurized at a point with jets of millions of gallons of high-pressure water, chemicals and sand mixture. Once the rock is fractured, shale oil gas is released. The USA has justified fracking projects with the assurance of a job boom. In 2018 the new Draft Five year Outer Continental Shelf Oil and Gas Leasing Program proposed opening of more than 90% of the US coast to fracking. In the UK more than 100 licenses have been issued for fracking in Lancashire, Yorkshire, Nottinghamshire and at Cuadrilla where serious legal battles have already started to question the practice. Marine and human life along the coasts and some prized outstanding marine national parks have been severely affected due to earthquake tremors, acidification and pollution of ocean and groundwater around the release openings. It is against this background that a small victory against fracking for animals such as whales, sharks and sea otters inhabiting a small area of the Pacific Ocean near the Californian coast should be

mentioned here. Their saviors were the Center for Biological Diversity, the Wishtoyo Foundation and the Environmental Defense Center who had filed three lawsuits against Donald Trump's government. This resulted in a landmark agreement of March 2019 toward better conservation practices in relation to coastal animals.

2. Whale shark conservation: Almost the entire coastline of India is inhabited by gentle sea giants called whale sharks. Since 2001, they have been pioneers for being the first sea species to be accorded the highest level of legal protection under Schedule I of the Wild Life (Protection) Act, 1972. The coasts of Kerala and Lakshadweep have faced serious threats due to the degradation of the marine ecosystem as a result of a variety of causes, of which coastal vegetation degradation tops the list, while policy-makers continue to debate the causes, such as pollution, climate change and unsustainable fishing practices. Vegetation has to be restored and other issues will be enforced in the process of implementing strict vegetation conservation practices along the coasts. However, an integrated project to collectively change the mindset of inhabitants along the coasts to stop overkilling and participate in conservation must be implemented. This has already been initiated through the Wildlife Trust of India's (WTI) award-winning Whale Shark Conservation Project with the fisherfolk of Kerala, Lakshadweep and Gujarat coast. The Whale Shark Conservation Project has been highly successful in changing the mindsets of fishing communities on the Gujarat coast. On August 30 2017, on the occasion of International Whale Shark Day, WTI announced that it was partnering with Cochin Shipyard Ltd. (CSL) to bring the project to Kerala and Lakshadweep and are now participating in tagging, monitoring and subsequently conserving the whale sharks. This is done through satellite information signals to forest conservators and other forest officials along the coastline in case the whale shark becomes entangled in fishing nets, which have become their major killers. Since 2017 many tagging exercises have proven to be a success.
3. Trash Free Seas: The Trash Free Seas program of the Ocean Conservancy Group is a dedicated effort to conserve marine species. Nicholas Mallos, the Director of this project states, 'Plastics now pollute all dimensions of our oceans from sea surface to the seafloor, on remote beaches and in the Arctic sea ice. The impact ocean plastics have on marine species is well documented, but increasingly scientists are concerned about the potential threat of plastics to species at the top of the marine food chain: humans'. Plastic Pollution

is the greatest man-made killer of ocean animals. More than 300 million tons of plastic waste is produced every year, which is equal to the weight of the entire human population (UNEP). Plastics are produced from chemicals derived from fossil fuels, which are non-renewable. It is estimated that by 2050 the plastic industry will end up using more than 20% of our fossil fuels and deposit more than 8 million tons of plastic into oceans, destroying marine life. Decision makers in coastal states are treating this as a routine problem even though it is now an existential issue. Scientific studies (Gall and Thompson 2015, pp. 170–179) have found that plastics are killing over 700 marine animals annually and pushing more than 17% of species to extinction. Researchers at Plymouth University found evidence of 44,000 animals and organisms becoming entangled in plastics or swallowing debris from across the globe. Plastics threatened almost 92% of sea animals already on the International Union for Conservation of Nature (IUCN) Red List. The Trash Free Seas program found plastic in the viscera of more than 59% of sea birds, 100% of turtle species and 25% of fish sampled for sea food markets.[8] UNEP[9] finds that 90% of these plastics are carried by rivers to the ocean, and therefore in order to make ocean life plastic free, action should be coordinated with river management authorities in countries from where rivers are entering seas.

4. Conservation of phytoplankton: Some good initiatives are being implemented under the Global Ocean Biodiversity Initiative (GOBI) to conserve phytoplankton or forests at the ocean floor on which marine ecosystems and animals rely for survival. Phytoplankton contribute 50–85% of oxygen to the earth's atmosphere and provide a base for several aquatic food webs.[10] They are not just food stores for whales, shrimp, snails and fish but also lungs for human beings (NOAA 2018). Sarah Watts (2017), writing for the *Pacific Standard*, quotes Andrew Barton, an oceanographer and researcher from Princeton University: 'The plankton absorb the carbon dioxide like a tree, and when they die, they sink to the bottom of the ocean and that carbon is locked away for thousands of years. If not for phytoplankton, the carbon dioxide (in the atmosphere) would be higher, and the climate would be warmer'. It is worrying that since 1950, 40% of these planktons have been destroyed and many marine spe-

cies have been almost wiped out, with many reduced to a dangerously low number. GOBI is an international partnership to conserve biological diversity in the deep seas and open oceans. It has also been working to develop data, tools and methodologies to identify ecologically significant areas, with an initial focus on the high seas and deep seabed beyond national jurisdiction. This initiative works in close collaboration with the German Federal Agency for Nature Conservation (BfN), IUCN, UNEP World Conservation Monitoring Centre, Marine Conservation Biology Institute, Census of Marine Life, Ocean Biogeographic Information System and Marine Geospatial Ecology Lab of Duke University. The initiative applies an ecosystem approach to conservation of marine species, as a result of which animals are being protected in their habitats. The problem is still so large that it may not be solved in the current framework. The phytoplankton move toward colder seas and as these seas become warmer this pushes phytoplankton further northward. Large marine animal life may not be able to move along with phytoplankton and those animals which do move may not be able to adapt to their new environment.

An Emergency Disaster Preparedness Plan for Coastal Wildlife Conservation

Regarding the ruthless fragmentation and destruction of wildlife spaces and habitats, the UNEP suggested a restorative plan in which overconsumption of wildlife (legal or illegal) should be completely prohibited. To achieve this, UNEP[11] has put forward a demand for commitment from governments. Broadly, this includes a commitment to maintain and conserve 'intact habitats' or prevent fragmentation of wildlife corridors. Another commitment is to conserve 'priority land', that is, fragile land areas. It also includes a commitment to 'cross-sectoral planning', which indicates a rejection of monolithic planning which arises from isolated and compartmentalized departments. Overall, what is demanded is a more ecological economy to address threats to wildlife.

The state of coastal wildlife degradation is a warning call for human inhabitants of the only planet we have. The UNEP's Marine and Coastal Ecosystem Well-Being Report (2006, p. 18) records:

- All seven species of sea turtles listed under the Convention on International Trade in Endangered Species (CITES). According to the IUCN Red List, three are critically endangered, three are endangered, and the status of the Australian flatback turtle (*Chelonia depressa*) is unknown [CT 19.2.2.1 and 19.2.2.2].
- The Atlantic grey whale and Caribbean monk seal have been driven to extinction. Many dolphins are threatened by bycatch [CT 19.2.2.2.].
- Globally, 91% of albatross species, 59% of penguins, 43% of shearwaters, and 40% of frigate birds are threatened [CT 19.2.2.3].
- Shorebirds are declining worldwide: of populations with a known trend, 48% are declining and only 16% are increasing. Overall 45 (34%) of African-Eurasian migratory shorebird populations are regarded as 'of conservation concern' due to their decreasing and/or small populations [CT 19.2.2.3].
- Of the shark, ray, and chimaera species assessed by IUCN,18% are listed as threatened, 19% near threatened, 37.5% data deficient, and 26% least concern [CT 4.4.2.2].
- On the coast. of southern California, the California mussel *Mytelus californianus* has become very rare, the ochre sea star is now almost never seen, the once abundant black abalone can no longer be found, and dozens of formally abundant nudibranch species are now rare.
- Some species of crocodiles are under threat of extinction, although none of the 23 known species has actually gone extinct.

The concern of coastal destruction reverberates in the World Economic Forum as well. The World Economic Forum's January 2019 meeting in Davos questioned whether the world leaders were sleepwalking into crisis. By saving species from natural disasters the world economy gains USD 300 billion. A similar sentiment was visible in the 2014 UNEP's Report 'A New Deal for Nature'. The health of oceans, seas and backwater animal life and vegetation has become a prime concern for achieving sustainable development goals. The problem as brought up in the UNEP's New Deal for Nature Report is that 47% of ecosystems have already been lost and the rate of recovery of the remaining moderately exploited ecosystems is around 1%.

The UNEP Report lays out a plan to protect coastal life and ecosystems. The focus is biodiversity, fisheries, habitat loss and wider aspects of sustainable development. It also suggests geo-engineering to nurture carbon sequestration alongside waste management services at source. Swift and sensitive action is needed for holistic aquaculture management intertwined with conservation of coastal wild life corridors and a contiguity of vegetation which is protected over the network of underground coastal aquifers and over the groundwater bodies, coastal rivers, estuaries and backwaters. Major administrative restructuring is required within the disaster management structures so that a vision of transdisciplinarity overshadows the existing vision of outdated silos of scientific determinism of magic pill policies.

Notes

1. Organic detritus is animal waste and dead and decaying plants and animals on the bottom of the pond. Decomposers, also known as detritovores, are bacteria and other organisms that break down detritus into material that can be used by primary producers, thus returning the detritus to the ecosystem. This maintains ecosystem balance and protects against many disasters.
2. National Remote Sensing Agency (NRSA), Hyderabad (1983). Mapping of forest covers in India from satellite imagery (1972–75 and 1980–82), summary report, pp. 5–6.
3. Government of India (1987). Mangrove in India. Status Report, Ministry of Environment & Forest, Government of India, pp. 1–150.and Kumar R (2000). Distribution of mangroves in Goa. *Indian Journal of Forestry* 23:360–365.
4. Jagtap et al. (1993) and Naskar (2004).
5. Rosenblatt et al. (2013) explain the roles of large top predators in coastal ecosystems.
6. Blue Carbon Initiative available at https://www.thebluecarboninitiative.org/about-blue-carbon
7. Jackson et al. (2012).
8. Ocean Conservancy, Fighting for Trash Free Seas, available at https://oceanconservancy.org/trash-free-seas/
9. Available at https://www.unenvironment.org/interactive/beat-plastic-pollution/ 2019.
10. Earth Sky (2015). How much do oceans add to world's oxygen? *Earth/Science Wire*, June 8, available at https://earthsky.org/earth/how-much-do-oceans-add-to-worlds-oxygen
11. UNEP (2014). A New Deal for Nature available at https://wedocs.unep.org/bitstream/handle/20.500.11822/28333/NewDeal.pdf?sequence=1&isAllowed=y

References

Atwood, T. B., & Hammill, E. (2018). The Importance of Marine Predators in the Provisioning of Ecosystem Services by Coastal Plant Communities. *Frontiers in Plant Sciences, 9*, 1289. https://doi.org/10.3389/fpls.2018.01289.

Catton, W.R., & Dunlap, R.E. (1978). Environmental Sociology: A New Paradigm. *The American Sociologist, 13*(1), 41–9.

DeMello, M. (2012). *Animals and Society: An Introduction to Human-Animal Studies.* New York: Columbia University Press.

Fackler, M. (2017, September 29). After the Tsunami, Sea Creatures from Japan Crossed an Ocean. *New York Times*, p. A9.

Gall, S. C., & Thompson, R. C. (2015). The Impact of Debris on Marine Life. *Marine Pollution Bulletin, 92*(1–2), 170–179.

Jackson, J. B. C., Kirby, M. X., Berger, W. H., Bjorndal, K. A., Botsford, L. W., Bourque, B. J., et al. (2012). Historical Overfishing and the Recent Collapse of Coastal Ecosystems. *Science, 293*, 629–638. https://doi.org/10.1126/science.1059199.

Jagtap, T. G., Chavan, V. S., & Untawale, A. G. (1993). Mangrove Ecosystems of India: A Need for Protection. *Ambio, 22*, 252–254.

Naskar, K. R. (2004). *Manual of Indian Mangroves.* New Delhi: Daya Publishing House.

NOAA. (2018). What Are Phytoplanktons? *National Ocean Service.* Website https://oceanservice.noaa.gov/facts/eutrophication.html. 25 June 2018

Peggs, K. (2012). *Animals and Sociology.* New York: Palgrave Macmillan.

Rahman, K. M. M., & Rakhimov, I. I. (2015). Ecological Effects of Oil Spill on Bangladesh Sundarbans Biodiversity. In *International Scientific and Practical Conference "Complex Problems of Technosphere Safety"* (Part. VI. 258p, 16–21). Voronezh: Voronezh State Technical University.

Rosenblatt, A. E., Heithaus, M. R., Mather, M. E., Matich, P., Nifong, J. C., Ripple, W. J., & Silliman, B. R. (2013). The Roles of Large Top Predators in Coastal Ecosystems: New Insights from Long Term Ecological Research. *Oceanography, 26*(3), 156–167. https://doi.org/10.5670/oceanog.2013.59.

Singh, A. (2016). Understanding Implementation Slippages of the CRZ Law, Coastal Ballads and Conservation Ironic. *Economic and Political Weekly, 51*(7), 13.

Singh, A. (2018). Rising Sea Spell Disaster, Discourse. *Deccan Chronicle*, p. 11.

Surrallés, A. (2017). Human Rights for Nonhumans? *HAU: Journal of Ethnographic Theory, 7*(3), 211–235. https://doi.org/10.14318/hau7.3.013.

The Spokesman Review. (2019, May 30, Thursday). Pigs Who Survived Flood Shot at Levee. *USA Today.*

UNEP. (2006). *Marine and Coastal Ecosystems and Human Well-Being: A Synthesis Report Based on the Findings of the Millennium Ecosystem Assessment*

(76pp). UNEP. Available at http://wedocs.unep.org/bitstream/handle/20.500.11822/18066/unep_2006_marine_and_coastal.pdf?sequence=1&isAllowed=y

Watts, S. (2017). Global Warming Is Putting the Ocean's Phytoplankton in Danger. *Pacific Standard.* Available at https://psmag.com/environment/global-warming-is-putting-phytoplankton-in-danger

York, R., & Longo, S. B. (2017). Animals in the World: A Materialistic Approach to Sociological Animal Studies. *Journal of Sociology, 53*(1), 32–46.

York, R., & Mancus, P. (2013). The Invisible Animal: Anthrozoology and Macrosociology. *Sociological Theory, 31*, 75–91.

Zedler, J. B., & Kercher, S. (2005). Wetland Resource Status, Trends, Ecosystem Services and Restorability. *Annual Review of Environmental Resources, 30*, 39–74. https://doi.org/10.1146/annurev.energy.30.050504.14424.

Amita Singh is Professor of Administrative Reforms and Emergency Governance at the Centre for the Study of Law and Governance, JNU, New Delhi, and founder chairperson and transdisciplinary scholar at the Special Centre for Disaster Research, JNU, New Delhi.

Protecting the Non-Human Animals of Coastal Ecosystems from Disasters

Surinder Verma and Shalini

Introduction

Disasters cause a huge loss of human and non-human lives and destruction of their ecosystems across regions. Such calamities are usually temporary in nature but create deep physical, mental and emotional distress for a longer period. Unfortunately, many millions of animals are affected every year because of severe catastrophes. Fritz (1961) highlights the fact that the losses take place in a geographical space and time, affecting individuals, groups, families, communities or institutions, as a result much of the loss which can be prevented is allowed to happen. Every society is disrupted by disasters yet resilience can be built faster where human and non-human[1] lives share mutual respect.

It has become important to understand the loopholes and deficits in law and governance while managing and understanding plans for

S. Verma (✉)
Delhi School of Social Work, Delhi University, New Delhi, India
e-mail: surind99_slg@jnu.ac.in

Shalini
National Institute of Educational Planning and Administration, New Delhi, India

A. Singh et al. (eds.), *Development in Coastal Zones and Disaster Management*, Disaster Research and Management Series on the Global South, https://doi.org/10.1007/978-981-15-4294-7_9

preparedness when it comes to non-human animals during and before the events of disaster. This chapter argues that the arguments for the management of non-human animals in disaster situations have somehow missed out in acknowledging their role in any ecosystem. Whether non-human animals are from coastal or from mainland areas, their intrinsic value and care during crisis situations demonstrate a total apathy in counting their needs in policies.

Further, this chapter discusses animals in the coastal zones of India, its unique ecosystem, their fragility during the disasters and disconnected rescue operations. This chapter also emphasizes the primordial relationship which exists between humans and non-humans as they share the same ecosystem for ages as a community. The laws of coastal zones such as the Coastal Regulatory Zone (CRZ) Notifications are looked into from the perspective of accommodating animals as they have a primordial right to exist beyond the charitable inclinations of human beings.

Primordial Right to Co-Exist

In India, animals, birds and sea creatures are considered to be an integral part of human society and they are mutually bonded. E. M. Foster has expressed a powerful presence of bonding between animals and humans in Indian culture. In Hinduism they are part of the daily worship. Apart from this, animals and birds symbolize ideas and concepts in Asian religions from India to China. In Hinduism particularly, a tiger is the companion of Goddess Durga, Shri Krishna is a cow protector, bull symbolizes power of strength on which Lord Shiva rides, elephant is seen as Lord Ganesha and monkeys as Hanuman; these animals are being venerated by Hindus in these forms. The intricate relationship humans share with the non-humans since ages provide emotional and economic resilience combined with an overall well-being (Smith 2012, p. 440). It has also been found that recovery from traumatic encounters of disasters is different for humans and non-humans and it needs further research from a sociological, psychological and economical standpoint (Provitolo et al. 2011, p. 48).

Over the decades there has been a large-scale attention given to disaster mitigation process. In fact, Disaster Risk Reduction (DRR) has been a common dimension used during disasters. Later DRR was replaced by a more holistic approach/process of RRM (Resistance building, Risk reduction and Mitigation) taking community as a whole. This process takes an account of the whole ecosystem, including humans, non-humans, the flora and fauna that together live in the same space since pre-historic times. Often holistic approach of RRM is not implemented or exercised efficiently owing to the fact that animals are also treated in hierarchies based on their contribution to economy as a result; particular species become more important than others. Speciesism is an area of thinking which analyses how one species becomes more important than the other. Waldau (2001) explains the mischief of the underlying idea of speciecism as an 'inclusion of all human animals within, and the exclusion of all other animals from, the moral circle' (p. 38).

As said by Thompson (2013), for decision makers human–animal relationship is much different than human–human relationship because they consider human life more valuable. In general, humans prefer to rescue humans first and animals are given low priority during any disaster (p. 125). This attitude or consciousness for saving humans first rather than non-humans is a self-conceited socio-psychological behaviour which influences disaster management planning. Worse still, companion animals amongst non-humans get a last priority in rescue operations (Singh 2015, 2016). In defence of this, Thompson (2013) states that human life is more valuable than non-humans (p. 126). The loss of humans can be acknowledged by the fact that it results in emotional stress, which is much traumatic than the psychological impact that is experienced on the loss of animals (companion animals) (Zottarelli 2010). Social thinking and policy dimensions are now changing as media depiction of animals during disasters forces the world to realize the need for their well-being too.

Coastal India: Forests and Animals

India has a coastline of 7516.6 km, which borders the Indian Ocean on the South, mainland and islands in Bay of Bengal, and the Arabian Sea on the West. Ten states and four union territories of India (Andhra Pradesh, Gujarat, Goa, Karnataka, Kerala, Maharashtra, Odisha Tamil Nadu, Telangana, West Bengal, Andaman & Nicobar Islands, Daman & Diu, Lakshadweep and Puducherry) share this coastline, with Gujarat having the longest coastline. There are 1382 offshore islands as well. Coastal regions and islands have large and dense forests. There is immense and also much unexplored diversity in the forest ecosystem which is being ignored in disaster management policies.

According to Champion and Seth (1968), following are the types of forests which the coastal regions and islands in India have:

- Tropical moist evergreen forests (Maharashtra, Karnataka, Andaman, Kerala and Tamil Nadu, West Bengal and Odisha)
- Tropical moist semi-evergreen forests (Andaman, Western Ghats, Maharashtra, Goa, Kerala, West Bengal and Odisha)
- Tropical moist deciduous forests (Andaman, Nicobar, Gujarat, Karnataka, Tamil Nadu, Kerala, West Bengal and Odisha)
- Littoral and swamp forests (Andaman and Nicobar Islands, Kerala and West Bengal)
- Sub-Tropical Broad-Leaved Hill Forests (Kerala, Tamil Nadu, Karnataka, Maharashtra, West Bengal and Odisha)
- Montane Wet Temperate Forests (Tamil Nadu, Kerala and West Bengal)

In the above types of forests there exists a rich variety of flora and fauna. However, disaster management institutions have neither in the past nor today ever planned to collaborate with the Forest Research Institutes, National Forest Academy, Animal Welfare Board of India, Animal Husbandry departments or various animal welfare organizations and NGOs to facilitate pre-disaster preparedness and post-disaster rescue and rehabilitation efforts in a sustained manner. The situation is equally precarious in terms of wildlife where disasters such as floods and tsunamis wipe off many unaccounted wild species. There is still no plan of state disaster management authorities on protecting zoo animals during any disaster like floods, earthquakes and hurricanes.

Coastal India: Coastal Zone Regulation (CRZ) Law

CRZ notification (Ministry of Environment, Forest and Climate Change, 1991, 2011, 2019) has been in operation since 1991 when it was brought out under the Environment Protection Act 1986. It applies to the conservation of vulnerable ecology and ecosystems, which contain water bodies such as rivers, estuaries, lagoons, creeks, coral reefs, mangroves, swamps and backwaters. These coastal areas were classified as CRZ-1, CRZ-2, CRZ-3, and CRZ-4 and the same was retained in CRZ 2003 notification as well. Supreme Court had directed the states to set up the state Coastal Zone Management Authorities (CZMA), which would be involved in the formulation of the coastal zone management plan (CZMP) so that implementation of coastal zone regulations can take place. Three judgements of the Hon'ble Supreme Court have paved a strong road for the conservation of coastal ecology. These cases are historic trend setters which gave relief to coastal animal life. The first case was *Indian Council for Enviro-legal Action v Union of India* [(1996) 5 SCC 281]. It highlighted that state governments should prepare their state-specific coastal zone management plans. In the second case, *S. Jagannath v Union of India* [(1997) 2 SCC 87], the court emphasized a need to protect the ecologically fragile coastal areas, safeguard the aesthetic qualities and uses of the sea coast while prohibiting shrimp aquaculture farms right on the sea coast. In the third case, *Vaamika Island v Union of India and Ors.* [(2013) 8 SCC 760], the court upheld the High Court's decision to demolish illegal structures.

The 2011 CRZ notification (Department of Environment and Forest 2011) safeguarded the livelihood of the fishing or other local communities which are directly or indirectly dependent on the coastal area for their livelihood. It included that the coastal land up to 500 m from the High Tide Line (HTL) and a stage of 100 m along banks of rivers, estuaries, creeks and backwater subject to tidal fluctuations is called the Coastal Regulation Zone (CRZ). It restricts the setting up of and expansion of any kind of industry operations or processes and manufacturing or handling or storage or disposal of hazardous substances, which can negatively affect the coastal zone. It also gave some special provisions to Goa, Greater Mumbai, Kerala and critically vulnerable coastal areas (CVCAs) like Gulf of Mannar (Tamil Nadu), Karwar and Kundapur (Karnataka), Chilka and Bhitarkanika (Odisha), Gulf of Khambat and Gulf of Kutch (Gujarat), Sunderbans mangrove area, Malwan (Maharashtra), Vembanad (Kerala), Coringa, East Godavari and Krishna Delta (Andhra Pradesh).

The 2019 notification of CRZ law replaces many of the conservation-oriented legal norms of 2011. It aims to promote greater sustainability based on scientific principles and protected by law. Ironically, there is no reduction in the Development Promotion Zone. Also, for CRZ-III (Rural) areas, it mentions two different categories: CRZ-III A and CRZ-III B. This new CRZ notification has also eased the Floor Space Index norms and has given permission for the development of more tourism infrastructure. In order to ease the development projects, it has streamlined the clearances. It also mentions that there will be no Development Zone of 20 meters for all islands.

There is also an economic viability argument taking a front seat and becoming quite visible in the 2019 notification. Sundarbans region of West Bengal and other ecologically sensitive areas which have been identified under Environment (Protection) Act, 1986, as Critical Vulnerable Coastal Areas (CVCAs) such as Gulf of Kutchh in Gujarat, Achra-Ratnagiri in Maharashtra, Gulf of Khambat, Vembanad in Kerala, Karwar and Coondapur in Karnataka, Gulf of Mannar in Tamil Nadu, Krishna in Andhra Pradesh and Bhaitarkanika in Odisha will now be managed with the involvement of coastal communities including fisher folk. This encourages stronger lobbies to devastate coastal ecology once again.

Phenomenal Yet Unaccounted Loss to Non-human Lives

It is notable that addressing loss of non-human lives in disaster management policies is mostly unexplained, as there is dearth of comprehensive data when it comes to animal casualties in India. However, various national and international conventions and policies such as India's National Disaster Management Plan (2018), Hyogo Disaster Mitigation Framework for Action (2005–2015), Sendai Framework for Disaster Risk Reduction (2015–2030) and many more have included non-human lives as an important part of their policies, but the outcomes does not give a satisfactory picture with regard to India (See Table 1). One of the prominent reasons for this is paucity of human resource.

Evaluating the difficulties encountered during disasters, it makes the argument stronger about improving preparedness, mitigation process and minimizing impact of disasters on animals. Although some disasters are associated with climate change, that is mostly exacerbated by human interference in environment (Preston et al. 2015, p. 728). Records of lives lost in the floods across India suggest that loss of non-human lives lost is confined to cattle herds only and other companion and wildlife animals lost are not well recorded across regions. Following is the table showing the number of cattle heads lost in the last five years (report prepared by the Ministry of Home Affairs) according to State-Wise Details of Damage Due to Cyclonic Storms/Heavy Rains/Floods/Landslides/Cloudburst and so on during the years 2013–14 to 2017–18:

Table 1 State-wise Details of Damage due to Cyclonic Storm/Flash Floods/Floods/Landslides/Cloudburst etc. during the years 2013–14 to 2017–18

Coastal States in India	*Number*	*Of*	*Cattle*	*Heads*	*Lost*
	2013–14	*2014–15*	*2015–16*	*2016–17*	*2017–18*[a]
Andhra Pradesh	2517	4777	3669	32	55
Gujarat	274	112	19,388	252	15,255
Goa	–	–	–	–	–
Karnataka	286	85	–	217	745
Kerala	1366	527	4	56	77
Maharashtra	2164	53	–	1135	164
Odisha	5688	672	–	–	399
Tamil Nadu	–	341	12,030	564	7654
Telangana	–	–	–	6535	–
West Bengal	45,285	145	23,120	3320	2075
Puducherry	48	–	1095	–	–
Lakshadweep	–	–	–	–	1656

Source: Annual Reports of Ministry of Home Affairs (2013–14, 2014–15, 2015–16, 2016–17, 2017–18)
[a]As on 31/12/2017

Major Disasters in Coastal Areas or Areas Regulated by CRZ Law of India: Presented as Small Cases

Few major disasters in India saw the loss of economy as related to the loss of human and non-human animal lives. Three disasters are being documented here as small cases to understand that there is dearth of specific provisions in the disaster- and environment-related laws; also, whatever is still in place is not being executed efficiently through the lens of RRM; this clearly shows that there is lack of coherence in theory and practice.

2019 Cyclone Fani

On 3 May 2019, the vicious super cyclone ripped through Odisha. In the first week of May 2019 Odisha experienced a category 4 storm, after twenty years according to records. Conditions on 3 May were marked by record or near-record strong surge, winds and extreme rainfall. The UNICEF India Cyclone Fani Situation Report 2 summarized the disaster as follows:

> *On 3rd May the cyclone Fani ("Foni"), a category 4 storm, impacted three states in India (Odisha, Andhra Pradesh and West Bengal). Odisha remains*

> *the worse hit with more than 15 million people incl. 4,8 million children affected. The cyclone was untimely; it was one of the rarest summer cyclones, the first one in 43 years and one of 3 to hit Odisha in the last 150 years.* (UNICEF 2019)[2]

Loss of lives was significant as reported in *Hindustan Times*:

> *Twenty years after the vicious Super Cyclone ripped through Odisha (then Orissa), killing more than 10,000 people and 200,000 animals, and destroyed physical infrastructure worth crores (state government estimates), Cyclone Fani, equivalent to a Category 4 hurricane (on the Saffir-Simpson Scale), made landfall at the temple town of Puri on May 3, causing damages in Bhubaneswar, Cuttack, Puri, and Khordha districts, killing 38 people in a state of 46 million.* (HT, May 2019)[3]

The Animal Welfare Trust Ekamra, an animal welfare organization in Odisha, while addressing the loss of non-human animals in disaster identified and acknowledged the significant problem in coordinating services to animals in emergency planning during and before arising of Fani. Animal Welfare Trust Ekamra reported that:

> *The cyclone intensity was much higher than expected. However, efforts were made to evacuate animals but it was hard to find the place to shift them during disaster. Some of the animals were shifted to small rooms but the intensity of the cyclone is so high, all the roofs* (Asbestos sheets roof) *were blown away … later all the animals were released in order to save them but post cyclone 15 animals were found paralysed and few were dead and some went missing.*[4]

2016 Assam Floods[5]

In mid-2016, the State of Assam was subject to serious flooding, resulting in loss of lives of human and non-human animals and significant property damage. Although, flood in Assam comes annually, in 2016 it caused a major disaster. The commission of enquiry presented a report and described Assam flood disaster in the following way:

- Prolonged and heavy rainfall in the state of Assam had affected million peoples and flooded the Kaziranga National Park. As per Assam State Disaster Management Authority report, nearly 17, 94,544 people were affected by the flood and approx. 1.25 lakh people took shelter in relief camps in Assam.

- More than 2 lakh hectare of crops were damaged in 23 districts due to breaching of the embankments of Brahmaputra.

Assam Disaster Management Authority presented a final flood report on 28 July 2016 and it was found that the report does not have any data related to the animals affected in national parks and wildlife sanctuary. However, 'Down to Earth' came up with the real picture and showed that there is lack of consideration for animals in prior emergency planning:

> *Nearly 250 animals, including 17 Rhinos and more than 166 Hog Deer died within a week after one of the worst floods in Assam led to the submergence of the Kaziranga National Park (KNP). The flood water also encroached upon the camps of forest guards. Between July 25 and July 31, the park also lost 11 Wild Boar, nine Swamp Deer, six Sambar, three Buffalo, two Hog Badger, one Porcupine and one Python. The death toll is likely to rise after the entire flood-water gets out of the park.*[6]

Down to Earth stated that the forest teams had managed to provide some help to the non-human animals. It further noted that:

> *Forest teams have managed to rescue at least nine rhinos, 90 Hog Deer, one Jungle Owlet, three Swamp Deer, one Fishing cat and one Python. The Rhinos are currently undergoing treatment at the Centre for Wildlife Rehabilitation and Conservation (CWRC), India's only wildlife field hospital in Kaziranga.*[7]

The World Wildlife Fund for Nature came up with a report that reflects on the gaps in existing rescuing systems, particularly for non-human animals. It states that the State Disaster Management Authority doesn't have any early warning system for flood warnings in Kaziranga National Park which is essential to avert animal deaths in future.[8] It further recommends that Kaziranga needs more highlands with good cover of herbs and grasses so that they can withstand severe floods in future and act as refuge for indigenous animals (ibid.).

2018 Kerala Disaster

The state of Kerala experienced an abnormally high rainfall starting from June till mid-August of 2018. This had resulted in severe flooding of almost all the districts of Kerala. As per IMD, Kerala received 2346.4 mm of rainfall from June to mid-August 2018, which is much higher than the expected rainfall Kerala receives on an average annually. Owing to heavy rainfall, Kerala got flooded towards the end of July. Almost all the reservoirs were full due to continuous rainfall and water was released from several dams, which resulted in flooding in 13 out of 14 districts of Kerala. In the final report of 2018 Kerala flood, the state government presented the data as follows:

> *According to latest reports of the state government, 1,259 out of 1,664 villages spread across its 14 districts were affected. The seven worst hit districts were Alappuzha, Ernakulam, Idukki, Kottayam, Pathanamthitha, Thrissur, and Wayanad, where the whole district was notified as flood affected. The devastating floods and landslides affected 5.4 million people, displaced 1.4 million people, and took 433 lives.* (22 May–29 August 2018)[9]

The loss of non-human lives was significant:

One of the BBC reports of 23 August 2018 mentioned that 'more than 8000 cattles, calves and buffaloes, 3297 goats and 47 dogs have died in the flood'.[10] Whereas another report on Kerala floods of 2018 by the Special Centre for Disaster Research (SCDR), Jawaharlal Nehru University (JNU), asserts that the loss of non-human lives are uncounted because the State Disaster Management Authority had neither undertaken pre-disaster mapping nor did the state government release any exact figure of animals lost in the floods. However, on-field surveys by the research team to the four worst-affected districts suggested a tentative loss of around 4 lakh birds and an equal number of companion pets and street animals (Singh et al. 2018).

To understand policy gaps in documenting data on animals, the study conducted by the Special Centre for Disaster Research (SCDR) and the UPE-2 Research team on Western Ghats, JNU drafted a final report. In one of the interviews given to the *Times of India*, Professor Amita Singh, Founder Chairperson of SCDR, JNU (Also see Singh 2019), spoke about Kerala disaster management operations which brought to focus an enormous philanthropy of local communities at the face of government failures.

> *The recent floods in Kerala reflected a total failure on the part of the State government. It was ill prepared to encounter a disaster. Although, ecologists had repeatedly warned the state government about its preparedness like management of dams, information dissemination to municipalities and panchayats authorities but State institutions on disaster management failed in alerting local bodies about the impending disaster.*[11]

Fig. 1 A Tweet from a public twitter account on rescue operations for companion animals from homes and streets

It demonstrated to the world how local communities risked themselves in one of the biggest and most dangerous rescue operations for companion house pets and homeless street animals (see Fig. 1). A deeper study, published in the Kerala Report (2018) by the research team after visiting homes of these animals rescued by Samaritans[12] (see photographs of some of his rescued animals in Fig. 2), has further strengthened Hyogo Declaration's pedagogical arguments insisting on community resilience building through better preparedness with local communities and their institutions.

These three cases unequivocally highlight major gaps in policy priorities where human concerns hardly leave any space for consideration of non-human animals. These cases also make a very strong case in support of community resilience building through community participation rather than focusing simply on rescue and relief operations which mostly become voluminous, politicized and unmanageable if communities are not involved and vulnerabilities mapped in advance preparations.

Fig. 2 Some of the rescued animals by a good community Samaritan Johnson V. Edicula in Upper Kuttunad village of Kerala's Alappuzha district

What Can Be Done?

The Constitution of India under Section 51A clearly specifies that 'It shall be the duty of every citizen of India; (g) to protect and improve the natural environment including forests, lakes, rivers, and wildlife, and to have compassion for living creatures'. In 2008 the National Disaster Management Guidelines on Management of Biological Disasters included a dedicated chapter for 'Management of Livestock during Disasters'. Following these guidelines, some states have even integrated animal protection into their disaster management plans and this sensitization led to an inclusion of provisions for 'Animal Care' in the Section 7.10.1 of the National Policy on Disaster Management 2009. It states that 'Animals both domestic as well as wild are exposed to the effects of natural and man-made disasters. It is necessary to devise appropriate measures to protect animals and find means to shelter and feed them during disasters and their aftermath, through a community effort, to the extent possible'.

There is no effective system developed so far by state governments and the disaster management authorities such as SDMAs or DDMAs to train volunteers for the rescue and smooth evacuation of different types of animals. *A primary need for any state disaster management authority is to undertake segregated mapping of all animals (big and small, domesticated, wild or homeless and abandoned), incorporate shelter homes and veterinary hospitals and animal experimentation institutes to get vulnerability mapping done and involve college and university students in training programmes for animal rescue and rehabilitation exercises.*

These gaps in disaster management compel us to believe that human life has remained the top priority and evacuation of non-human animals has not been incorporated widely in disaster planning despite revealing studies on the similarities of behaviour between human and nonhuman animals (Tanaka et al. 2019). Most communities are also not readily involved in mitigating animal suffering during disasters for lack of proper training to deal with emergencies (Heath and Linnabary 2015, p. 183).

This issue has also been raised by one of the animal protection bodies, World Animal Protection (WAP) on 13 October 2018 on the occasion of International Day for Disaster Mitigation. They asked the Government of India to develop Animal Disaster Management plan with all financial and technical resources to protect animals from any disaster in near future.[13] Keeping aside an economic fact that so many communities depend for their livelihood on animals, it is also an issue of sentience and ecological justice. It is an indispensable need for any civilized society to understand that the web of life is a combination of innumerable ecosystems of life forms which keep together every human and non-human animal. Ironically, in all CRZ notifications including the latest notification of 2019, there is nothing related to saving non-human lives in terms of protecting them from the destruction during disasters. The notification of 2019 clearly emphasizes business, commerce and relief to humans inhabiting the coasts despite serious impact these activities have been having on ocean life, from corals to whales and phytoplanktons at the ocean floor:

> *In the CRZ areas, the fishing villages, common properties of the fishermen communities, fishing jetties, ice plants, fish drying platforms or areas infrastructure facilities of fishing and local communities such as dispensaries, roads, schools, and the like, shall be indicated on the cadastral scale maps. States and Union territories shall prepare detailed plans for long term housing needs of coastal fisher communities in view of expansion and other needs, provisions of basic services including sanitation, safety, and disaster preparedness.*[14]

This chapter suggests that every village should setup animal shelters, food, medicines and fodder storages for animals, floats and boats. Most important of all is the task of **'carrying forward a continuous Animal Birth Control Program led by an alert Municipality or a local Panchayat in collaboration with local animal welfare groups and people irrespective of a wait for the disaster to strike'.** This is best done with the help of communities and other civil society volunteers. Volunteers could be trained alongside civil society-based organizations, veterinary institutes, animal husbandry departments, SDMAs and district-level disaster management authorities. A separate fund should be earmarked to facilitate quick action on the above.

*A committed and intricately monitored census of animals including pet, dairy, urban, forest and wild animals should be undertake*n. It should be done by the SDMAs with the help of district-level authorities such as animal husbandry and civil society volunteers. Proper record keeping will ease the process during and after disasters. *Last of all, there should strict penalty for the State-, district- or village-level authorities for their inefficiency, dereliction of duty and negligence towards sentient beings.* Kautilya writes in his most famous book *Arthashastra*, "*Punishment is extremely important because a person who committed crime but was not punished is similar to a big fish that eats a tiny one, and powerful people will start hurting weak people. Punishment provides strength to the weak people*" Sharma (2018, p. 59). As discussed above, the government should come up with a comprehensive and robust plan to protect animals during disasters.

Conclusion

This chapter has outlined the importance of all sentient beings or animals as inhabitants of planet earth. Various gaps in the planning, preparedness, awareness and resilience building of non-human animals would be eradicated as one is able to put forth a well-researched ecosystem management approach at the centre of all sciences and social sciences.

Notes

1. 'Non-humans' has been used for all kinds of animals in this chapter.
2. Accessed on 15 June 2019 from, https://reliefweb.int/report/india/unicef-india-cyclone-fani-situation-report-2-national

3. Accessed on 15 June 2019 from, https://www.hindustantimes.com/editorials/cyclone-fani-what-india-can-learn-from-odisha/story-TJmZ5cRH4IxUZIWIkjqZLM.html
4. Animal Welfare Trust Ekamra. Submission to *Business Standard*, 15 May 2019. Available online: https://www.business-standard.com/article/news-ani/cyclone-fani-animal-lovers-seek-help-to-rebuild-shelter-homes-in-odisha-119051500529_1.html (accessed on 15 June 2019).
5. Although Assam doesn't come in coastal zone of India but CRZ law applies in Assam as well.
6. Down to Earth published a report on 2 August 2016, Available online: https://www.downtoearth.org.in/news/wildlif-&-biodiversity/amp/assam-floods-about-250-animals-including-17-rhino-dead-in-kaziranga-55133 (accessed on 15 June 2019).
7. Down to Earth publish a report on 2 August 2016, Available online: https://www.downtoearth.org.in/news/wildlif-&-biodiversity/amp/assam-floods-about-250-animals-including-17-rhino-dead-in-kaziranga-55133 (accessed on 15 June 2019).
8. Down to Earth published a report on 2 August 2016, Available online: https://www.downtoearth.org.in/news/wildlif-&-biodiversity/amp/assam-floods-about-250-animals-including-17-rhino-dead-in-kaziranga-55133 (accessed on 15 June 2019).
9. Government of Kerala published a report on October 2018, Available online: https://www.undp.org/content/dam/undp/library/Climate%20and%20Disaster%20Resilience/PDNA/PDNA_Kerala_India.pdf (accessed on 15 June 2019).
10. Available online: https://www.bbc.com/news/world-asia-india-45267644(accessed on 15 June 2019).
11. *Times of India*, 22 January 2019, Available online: https://timesofindia.indiatimes.com/city/kochi/state-failed-in-disaster-mgmt-amita-singh/articleshow/67631687.cms
12. Mr. Johnson V. Edicula in Upper Kuttunad village in Kerala's Alappuzha district rushed to save pets from houses drowning in water as most residents were out for work in the city. He was later awarded with a good citizenry and Social Worker medal by many animal organizations across the world as well as a Church in the state.
13. Available online: https://www.dailypioneer.com/2018/sunday-edition/%2D%2D-include-animals-in-disaster-management-plans%2D%2D-.html (Accessed on 15 May 2019).
14. Department of Environment, Forest and climate Change, CRZ 2019 Notification opens the coast to more development and less protection.

References

B.B.C. (2018). *India's 'Biggest' Pet Rescue Operation in Kerala Floods.* Retrieved from https://www.bbc.com/news/world-asia-india-45267644

Business Standard. (2019). *Cyclone Fani: Animal Lovers Seek Help to Rebuild Shelter Homes in Odisha.* Retrieved from https://www.business-standard.com/article/news-ani/cyclone-fani-animal-lovers-seek-help-to-rebuild-shelter-homes-in-odisha-119051500529_1.html

Champion, S. H., & Seth, S. K. (1968). *A Revised Survey of the Forest Types of India.* New Delhi: Manager of Publications.

Fritz, C. E. (1961). Disasters. In R. K. Merton & R. A. Nisbet (Eds.), *Contemporary Social Problems. An Introduction to the Sociology of Deviant Behaviour and Social Disorganization* (pp. 651–694). Riverside: University of California Press.

Goswami, S. (2016, August). Assam Floods: About 250 Animals, Including 17 Rhinos Dead in Kaziranga. *DownToEarth.* Retrieved from https://www.downtoearth.org.in/news/wildlif-&-biodiversity/amp/assam-floods-about-250-animals-including-17-rhino-dead-in-kaziranga-55133

Government of Kerala. (2018). *Kerala Post Disaster Needs Assessment Floods and Landslides – August 2018.* Retrieved from https://www.undp.org/content/dam/undp/library/Climate%20and%20Disaster%20Resilience/PDNA/PDNA_Kerala_India.pdf

Heath, S. E., & Linnabary, R. D. (2015). Challenges of Managing Animals in Disasters in the US. *Animals, 5*(2), 173–192.

Hindustan Times. (2019). *Cyclone Fani: What India Can Learn from Odisha.* Retrieved from https://www.hindustantimes.com/editorials/cyclone-fani-what-india-can-learn-from-odisha/story-TJmZ5cRH4IxUZIWIkjqZLM.html

Ministry of Environment, Forest and Climate Change. (2019). *Notification.* Retrieved from http://www.ncscn.res.in/pdf_docs/crz-2019.pdf

Preston, J., Chadderton, C., Kitagawa, K., & Edmonds, C. (2015). Community Response in Disasters: An Ecological Learning Framework. *International Journal of Lifelong Education, 34*(6), 727–753.

Provitolo, D., Dubos-Paillard, E., & Muller, J. (2011). *Emergent Human Behavior During a Disaster: Thematic Versus Complex Systems Approach.* Paper presented at EPNACS (Emergent Properties in Natural and Artificial Complex Systems) Sept. 15, 2011, Vienna, Austria, available at https://www.researchgate.net/publication/.

Sharma, R. K. (2018). Disasters: Ubiquitous Legal Framework in Ancient BC literature. In A. Singh (Ed.), *Disaster Law: Emerging Thresholds.* Taylor & Francis.

Singh, A. (2015). Disaster Management Is Not Only About Humans. *Hindustan Times*, July 25th, 02:29 IST.

Singh, A. (2016). The Centre's Decision on Culling Flouts Its Own Internal Findings. *Hindustan Times*, June 28th, 00:18 IST.

Singh, A. (2019). *State Failed in Disaster Management.* Retrieved from https://timesofindia.indiatimes.com/city/kochi-in-disaster-mgmt-amita-singh/articleshow/67631687.cms

Singh, A., Chugh, G., Kamthan, M., et al. (2018). *August 2018 Floods in Kerala.* A Report on Governance and Environmental Management, UPE 2 Project on Western Ghats, Jawaharlal Nehru University, Delhi (Departmental Publication).

Smith, B. B. (2012). The 'Pet Effect': Health-Related Aspects of Companion Animal Ownership. *Australian Family Physician, 41*(6), 439–442.

Tanaka, A., Saeki, J., Hayama, S. I., & Kass, P. H. (2019). Effect of Pets on Human Behavior and Stress in Disaster. *Frontiers in veterinary science, 6.*

Thompson, K. (2013). Save Me, Save My Dog: Increasing Natural Disaster Preparedness and Survival by Addressing Human-Animal Relationships. *Australian Journal of Communication, 40*(1), 123.

UNICEF. (2019). *India Cyclone Fani Situation Report 2 National.* Retrieved from https://reliefweb.int/report/india/unicef-india-cyclone-fani-situation-report-2-national

Waldau, P. (2001). *The Specter of Speciesism: Buddhist and Christian Views of Animals.* Oxford: Oxford University Press.

Zottarelli, L. K. (2010). Broken Bond: An Exploration of Human Factors Associated with Companion Animal Loss During Hurricane Katrina. *Sociological Forum, 25*(1), 110–122. https://doi.org/10.1111/j.1573-7861.2009.01159.x.

Surinder Verma is an assistant professor at the Delhi School of Social Work, University of Delhi.

Shalini is a doctoral scholar at the National Institute of Educational Planning and Administration, New Delhi.

PART IV

Tackling Vulnerability and Resilience in Coastal Ecosystems

Building Resilience in Coastal Ecosystems: Problems and Prospects

Akanchha Singh

Introduction

Disasters are unforeseen events which test the endurance and resilience of nature and individuals alike. Resilience implies acknowledgment of the inevitability of change, yet remaining unchanged by the consequences thereof. It is the ability to live with change and uncertainty without being impacted detrimentally, and to learn from past experiences and respond better to future disturbances.

Resistance and resilience are related concepts but with some difference. Resistance refers to the ability of natural systems to withstand changes without modification, whereas 'resilience' measures the ability of the system to respond to disturbances and regain their original equilibrium. Therefore, resistance as a concept has a pre-disaster context while resilience features in the post-disaster scenario. Thus, in the long run, resilience is a desirable attribute.

A. Singh (✉)
Jawaharlal Nehru University, New Delhi, India

A. Singh et al. (eds.), *Development in Coastal Zones and Disaster Management*, Disaster Research and Management Series on the Global South, https://doi.org/10.1007/978-981-15-4294-7_10

Coastal Ecosystems

Coastal ecosystems are inherently vulnerable, being exposed to the forces of seas and oceans while braving internal disturbances. The impact of climate change is more profound on these ecosystems. Sea level rise, salt water intrusion, increase in sea surface temperature, ocean acidification, coastal erosion and flooding are some significant concerns in marine ecology. Of late, human activities have exerted pressure on natural ecosystems which threaten their resistance and resilience.

Given the existence of these hazards, it is imperative to build internal resilience to withstand the impact of disasters.

Simonovic and Peck (2013) have identified four intertwined areas that determine coastal resilience. These are natural environment, built infrastructure, governance networks (institutions and interactions) and social dynamics. For instance, the location of an infrastructure determines its vulnerability to extreme waves in coastal set up. Thus, large and significant infrastructure should be placed either inland or at a higher elevation so as to withstand ferocious waves and storm surges. Further there are also examples to show that vegetation helps control coastal erosion.

Recently, Rosati et al. (2015) have stated that coastal resilience is based on four concepts: prepare, resist, recover and adapt.

Multi-Hazard Prone Coastal Systems

Coastal destinations are a hub for tourism, trade and other lucrative economic opportunities. Thus, there is a temptation to develop these areas in terms of infrastructure and services often overlooking the environmental costs of such an exercise. Natural hazards translate into catastrophic disasters as a consequence of these developmental projects threatening the stability of coastal ecosystems.

Coastal zones are used for different and sometimes competing purposes such as waste disposal, habitation, harbors, polluting industries, agriculture, tourism and recreational activities. Of late, there seems to be aggressiveness about maximizing output from these economic activities, in the process detrimentally impacting the sensitive coastal ecosystem. Biological productivity is fast declining; corals near the shore are becoming progressively bleached and inter-tidal habitats are disappearing by the day. Thus, there is a need to re-examine how coastal zones are managed.

Another concern with coastal regions is that they are subject to recurrent hazards; cyclones and floods are annual episodes in these areas. Thus, the need to be constantly prepared for the unforeseen cannot be overemphasized. The community approach to disaster management is important to examine against this backdrop.

India- Where Do We Stand?

There has been a sincere effort to regulate activities in coastal zones in order to mitigate the impact of impending disasters. This includes resource management, putting environmental norms in place and prohibiting selected activities. Coastal Regulation zone norms were issued by Ministry of Environment and Forest. Now, there is discussion about integrated coastal zone management using an inter-sect oral approach. However, the success of plans lies in its effective implementation. In practice, the top-down regulatory approach has diluted the intention behind such plans.

The draft Coastal Regulation Zone (CRZ), 2018, is set to replace the 2011 regulations. The draft bill has the potential to change the way in which coastal regions are governed. The new regulations aim to boost tourism in coastal areas, the potential of which has remained untapped due to a stringent regulatory framework. Critics argue that the proposed changes will simultaneously promote commercialization.

There is a need to review the 2011 regulations taking different stakeholders on board. Also, it is important to understand that coastal areas have their own dynamics which are quite distinct from the inland areas. Thus, policy makers should rely on both theoretical studies as well as practical insights from the field while formulating regulations for these vulnerable zones.

Building Resilience in Coastal Cities

Coastal areas have been historically overpopulated. India has a 7500 km-long coastline along which major hubs of settlements grew due to trade linkages; Kolkata, Mumbai, Chennai are some examples as well as numerous towns such as Vishakhapatnam, Masulipattanam, Vadodara, Mangalore and Puri. Major disasters have hit the Indian subcontinent on both the eastern and western coasts, including annual cyclones, tsunamis and coastal flooding.

Because cities have their own unique problems and requirements, Disaster Management frameworks have to be customized to accommodate their special needs.

Lacunae in Disaster Planning and Solutions Thereof

a) Needs of vulnerable population

An absence of ward-level data on vulnerable populations (the elderly, children, pregnant women and differently abled) is another major problem. Such data is vital for any relief and rescue operation in the aftermath of a disaster.

There is a need for separate provisions for the differently abled. When a warning is issued hours before a disaster is about to strike, there is complete chaos. In this situation people with special needs are left to fend for themselves. Amid the chaos they tend to lose valuable devices such as hearing aids and crutches, which leaves them more vulnerable than ever.

A disaster is usually seen as a leveler, in terms of its effect on people of all social classes. However, it is still a great deal harder on people with disabilities, who have, in the past, lobbied for better accessibility in cities. Activists and people with disabilities alike hope that the increasing intensity and frequency of human-induced disasters will jolt the government into finally taking their cause more seriously.

NGOs and other voluntary groups can provide help to people with special needs.

b) Medical facilities

There is a need for an efficient network of emergency care that has an assured provision of life support ambulances linked to trauma relief centers.

It is usually the case that hospitals are unprepared to tackle such emergencies. Underequipped in terms of infrastructure (beds, medical devices used for diagnosis) and plagued by staff shortages, hospitals resort to improvisation as the situation worsens.

Many private hospitals providing expensive medical care have emerged as an alternative to the largely defunct public hospitals. These cater to a large affluent middle- and upper-class population. However, they flout all safety norms and aim to carve out a business opportunity from the crisis.

c) Urban planning

Poor land-use planning, indiscriminate approval of building plans (often in violation of environment and municipal by-laws), and the total absence of disaster-risk assessment in urban design have resulted in what experts term 'concentrated concretization', which predisposes cities to disaster risks. Abating disaster risk requires a mainstreaming of disaster planning. To do this, the effective participation of nodal agencies–the Urban Local Body, whether Municipal Corporation or Town Council–is key. For instance, in case of floods Urban Local Bodies are at the heart of all operations.

Reasonable legislative autonomy and devolution of financial powers is indispensable to urban resilience. Cities are rightly seen as the engines of economic growth. In the case of India, cities with roughly a third of the total national population contribute to more than 63% of the GDP. Against this backdrop, municipal revenues account for 0.75% of India's GDP (which is meager compared to other developing countries).

Disaster Preparedness in 'Smart Cities'

McKinsey reports that by 2030 the number of million plus cities is expected to expand to 80 from the present-day 53, adding 600 million more people. As urban disasters are inevitable, this would render more people vulnerable.

In addition, the concentration of people in risk-prone areas such as flood plains, coasts and the Himalayan belt results in aggravation of the impact that a disaster would otherwise have. Such a situation necessitates a cost—benefit analysis of disaster risk reduction.

Smart cities are supposed to be technology driven with a stress on state-of-the art infrastructure. It would therefore make sense to build transport, communication, water, sanitation and power infrastructure with optimum physical resilience.

Identifying and Mitigating Risk

The first step in building a resilient city is to identify risks both at the micro and the macro level. Second, it is important to identify the vulnerabilities of communities and potential exposure to disasters. Third, an urban risk assessment exercise needs to be undertaken to identify critical infrastructure and develop early warning systems.

Mitigating risks calls for developing both structural and non-structural measures. While structural measures would include dams, wave barriers and retrofitting of buildings, the non-structural measures comprise policies and laws, practices and agreements such as building codes, land-use planning, public awareness and information, which aim at moderating the impacts of disasters.

Strict laws need to be implemented against substandard construction practices which are being rampantly carried out to save on marginal cost.

In the aftermath of a disaster, care must be taken that infrastructure is not built in hazard-prone areas. This helps in arresting additional risk.

Building Blocks of Resilience

a) Diversifying livelihoods

Coastal ecosystems are characterized by a range of livelihood opportunities such as fishing, boat making, foraging, agriculture, poultry, aquaculture, livestock ranching and employment in the burgeoning tourism industry. The wider the livelihood portfolio, the greater will be the resilience. As the nature of economic activities changes with time, it should be noted that the use of technology should be in sync with sustainable utilization of resources.

b) Governance

The best of laws on paper are hollow without effective implementation. Though established by-laws and environmental regulations may be listed, their effective implementation must be ensured. It was observed in the aftermath of tsunami 2004 that houses were built very close to both the eastern and western coasts (less than 20 meters). Moreover, poor building material was used which could not withstand the force of strong waves.

Empirical evidence suggests that Coastal Regulation norms are honored more in breach. Therefore, responsible, accountable and transparent governance is a key factor in building local resilience.

c) Resource management

Decentralized resource management is a strong pillar of community resilience. It includes efficient management of local resources and livelihood such that it reduces risk of natural hazards. This may include the conservation of critical habitats such as corals reefs and wetlands or creation of bio-shields such as mangrove barriers to withstand waves.

d) Augmenting social capital
Building trust and goodwill between people will go a long way in building community resilience. These social bonds can be leveraged in times of need. In India, people primarily rely upon social support systems such as family or neighbors in case of a disaster. But in fast-developing urban centers, family members may be separated by transnational boundaries, and neighbors, despite years of coexistence, may remain complete strangers. Thus, during disasters, when electricity, telephone and internet connections fail it becomes impossible to seek help.

The maximum brunt of disasters is borne by senior citizens; they form a major proportion of total deaths in case of a disaster. Among the senior citizens, the most vulnerable are those who are immobile or completely bedridden. Such people are left at the mercy of other family members. However, most of the young population go out for work during the day time leaving the elderly unattended, and in this situation, their already high vulnerability is accentuated. Such people are left defenseless and hopeless in the face of a disaster.

e) Mobilizing resources and nurturing local-level institutions to respond to emergency situations (micro regional planning)

This entails a bottom-up approach to planning rather than the bureaucratic, externally imposed plan of action. It takes into account participatory management of community resources, identification and mobilization of local resource personnel, keeping an inventory of commodities and grassroots-level capacity building initiatives. It would ensure equity in access as well as allocation of resources.

f) Capacity building (developing coping strategies)
Frequent mock drills and awareness campaigns help prepare communities for such extreme eventualities. An attempt should be made by state and non-state actors to develop coping capacities for communities to deal with disasters. Information on preparedness can be disseminated through Self Help Groups (SHGs), NGOs or student groups via group discussions, skits, door-to-door campaigns or the use of mass media. Involving local communities is critical in disaster preparedness. The community as the first responder in case of a disaster needs to have skills and training to deliver the best response in times of need. In Odisha, for example, local volunteers have been trained and equipped to provide first-aid and to conduct search and rescue operations, with special evacuation procedures to be followed for the disabled and elderly.

The recent Kerala floods in 2018 also brought to light how fishermen reached out to the far-flung flood-battered region, extending relief. Braving tough weather and rising water levels, hundreds of fishermen set out to be emissaries of relief and reconstruction work in Kerala. They reached out to those areas where the National Disaster Response Force could go and airlifting was not an option, especially in Chenngannur and Kuttanad. They were hailed, and rightly so, as 'sons of the sea' by survivors of the deluge. According to news reports, roughly 4500 fishermen with 1000 boats rescued nearly 65,000 people in districts such as Pathanimattha, Alappuzha, Thrissur and Ernakulam within the brief span of four days. After the situation stabilized, these fishermen reported that they had participated in the rescue work as they knew the terrain better as they depended on it for their livelihood. In addition, the bond of fraternity which they shared with fellow Keralites was the key driving force. This case study reflects how coping strategies along with social capital can be instrumental in relief and rescue work post disaster.

g) Using social memory as a source of innovation (by combining different kinds of knowledge)

Local traditional knowledge and experience is significant in understanding the dynamics of any ecosystem. Their inclusion in resource management and monitoring could complement state-of-the art scientific knowhow and lead to better outcomes.

h) Use of technology to build resilience

With improvement in remote sensing capabilities, it is easier now to monitor coastal ecosystems. Marine Remote Information Services (MARSIS) is now useful in mapping changes to the coastline, vegetation, level of erosion, inland water intrusion and primary productivity of corals. Sea Surface Temperature can also be measured by thermal sensors. Recently, satellite data has also been used to demarcate potential fishing zones. Modern-day state-of-the art knowledge should be supplemented with traditional coping strategies for best outcomes.

Community-Based Disaster Management in Coastal Ecosystems

Sushma Guleria and J. K. Patterson identify eight elements of resilience in coastal communities: Governance, Coastal Resource Management, Land Use, Structural Design, Society & Economy, Risk Knowledge, Warning and Evacuation, and Emergency Response and Disaster Recovery.

In addition, there is increasing advocacy for community-based disaster preparedness frameworks as they are the first responders in case of any emergency. Countries which are frequently subjected to natural disasters have a community-driven management framework which capitalizes on local resources and capacities to reduce vulnerability. The Philippines, for example, has had a Citizen Disaster Response Network since 1984, to carry out citizenry-based disaster management. As success stories about community-based disaster management come to light, more NGOs and government agencies across the world have adopted this as the standard modus operandi. There is a need to learn from experiences across the world and adopt their best practices for disaster preparedness and mitigation.

The aim of community-based disaster management is vulnerability reduction by augmenting the coping strategies of individuals to withstand disasters. Characteristic features of community-based disaster management include people's participation, acknowledgment of traditional coping strategies and prioritizing the needs of vulnerable groups. Contrary to popular assumptions, the role of community does not lie outside the government's structural framework, but is integral to it. The government should help by institutionalizing the framework of community-based disaster response mechanisms.

Problem of Quantification

A practical problem with the concept of resilience is that it cannot be easily quantified. While some factors which contribute to resilience can be captured by numbers, most others defy enumeration. Thus, a mix of quantitative and qualitative variables taken together would better explain the resilience quotient.

The notion of 'resilience' is widely accepted among scientists and environmentalists alike. However, when it comes to operationalizing the concept, there seems to be a lack of consensus. There is a disconnect between the theoretical understanding and practical application of what 'resilience' entails. World-wide, coastal management plans have incorporated the notion of resilience, either explicitly or implicitly.

Conclusion

Disasters, major or minor, remind us that improper planning and dense population together are a deadly cocktail for a catastrophe in coastal areas. Therefore, today, as India revamps its infrastructure, it is time to build resilience into the blueprint for future development.

At the global level, too, efforts to boost urban resilience are gaining momentum.

Lessons from across the world show that much can be done to reduce risk, even in complex coastal scenarios. To increase resilience, critical infrastructure and services such as schools, hospitals, water, electricity, communications systems and transportation will need to be built or retrofitted to withstand multiple hazards so that they continue to function in an emergency. Disaster Management should be mainstreamed in the development planning discourse to build future smart and resilient settlements.

Various organizations working in the field of Disaster Management have hailed the need for a paradigm shift. There is a need to be proactive rather than reactive when it comes to managing disaster risk. Thus, given the significant role played by coastal ecosystems, added to the increasing vulnerability that the people inhabiting these areas face, there is a need to come up with a separate coastal risk management framework.

References

Rosati, J.D., Touzinsky, K.F., & Lillycrop, W.J. (2015). Quantifying Coastal System Resilience for the US Army Corps of Engineers. *Environment Systems and Decisions, 35*(2), 196–208.

Simonovic, S.P., & Peck, A. (2013). Dynamic Resilience to Climate Change Caused Natural Disasters in Coastal Megacities Quantification Framework. *British Journal of Environment and Climate Change, 3*, 378–401.

Akanchha Singh is a research fellow at the Centre for the Study of Regional Development (CSRD), Jawaharlal Nehru University. She is studying the 'Political Economy of Post-Disaster Reconstruction', rooted in the critical realist paradigm which views reality as multi-layered. She has been working in the area of disaster management for five years, beginning 2013. She has been closely associated with the National Institute of Disaster Management, National Disaster Response Force, Sphere India, Caritas India and Red Cross Society while undertaking various research projects. She also got the chance to deliver a keynote presentation on the 'Role of Youth in Disaster Risk Reduction for South Asia' (December 10, 2014), at SAARC Disaster Management Centre, New Delhi, to member delegates of SAARC countries based on Delhi University Innovation Project findings.

Singh has been trained in the discipline of Geography, being a University Rank Holder at both graduate (Miranda House, University of Delhi) and postgraduate (CSRD, Jawaharlal Nehru University) level. She is the recipient of a UGC Postgraduate Merit Scholarship for University Rank Holder (2015–17) and a UGC Junior Research Fellowship (2018–2023).

Sustainable Development Goals (SDGs) and Risks to Coastal Communities

Sushma Guleria

There is an urgent need to review and integrate all relevant concerns out of the total 17 Sustainable Development Goals (SDGs), prepare management plans to pre-empt decision makers, and seek policy interventions to develop better public outreach programs. These concerns must empower coastal agrarian communities to plan strategies to improve and maintain their resilience to natural hazards. For example, Coastal Community Resilience (CCR) assessment results can be used to develop local actions, and to improvise local, sub-national and national plans, and can be met through high standards of governance at all levels.

The United Nations Sustainable Development Summit held on September 25, 2015 witnessed the adoption of an honest attempt at securing safe living. The SDGs substantially reflect India's development agenda. Even though the struggle to access basic service essentials is ongoing, India remains committed to protecting the environment alongside development with genuine intent, and hence is marching steadily yet diligently

S. Guleria (✉)
National Institute of Disaster Management, Ministry of Home Affairs, Government of India, New Delhi, India
e-mail: sushma.nidm@nic.in

A. Singh et al. (eds.), *Development in Coastal Zones and Disaster Management*, Disaster Research and Management Series on the Global South, https://doi.org/10.1007/978-981-15-4294-7_11

toward achieving the duly ratified SDGs. The 17 SDGs, which involve 169 targets, encompass scope for working toward resilience building. In addition, all of these are interrelated and the accomplishment of one will invariably and imperatively require tackling the others.

Increasing Vulnerability of Coastal Communities

Coastal areas globally are witness to growing disaster risks. Conspicuous impacts are seen in terms of surge in extreme weather events, sea level rise, coastal flooding and erosion, ocean acidification, glacial retreat, land and forest degradation, and loss of biodiversity and desertification. Additionally, unplanned development in the form of informal settlements, overcrowding, imprudent farming practices and inadequate infrastructure is proving to be a vulnerability amplifier. Over half of all coastal areas are urbanized and 21 of the world's 33 mega cities lie in coastal flood zones. The 7500 km-long Indian coast is occupied by nearly 20% of its total population and is uniquely characterized by varied landforms and ecosystems with enriched biological productivity and diversity, thereby attracting human settlements and commercial activities. These coastal products and services support human life and thus add to the nation's socio-economic progress card, corroborating the need to ensure their robust health through sustainable management, such that they continue to provide various goods and services for future generations.

Coastal Community Resilience is the capacity of a community to adapt to and influence the course of physical, environmental, social and economic changes. Such assessment of resilience and risks can be useful to characterize the resilience status and trends at the community level and can identify strengths, weaknesses and gaps in resilience building. In all, eight elements of resilience have been identified which are considered essential to reduce risk from coastal hazards, accelerate recovery and adapt to changing conditions in any affected community. These dimensions of resilience building can be analyzed and assessed through semi-structured interviews, focus group discussions or gathering information from key informants, which includes discussion with people who can provide specific valuable information such as those who have faced previous disasters. Moreover, these elements are intricately linked to the SDGs of water, health and poverty alleviation, which are critical to empowering communities. There are other targets in agriculture and rural development,

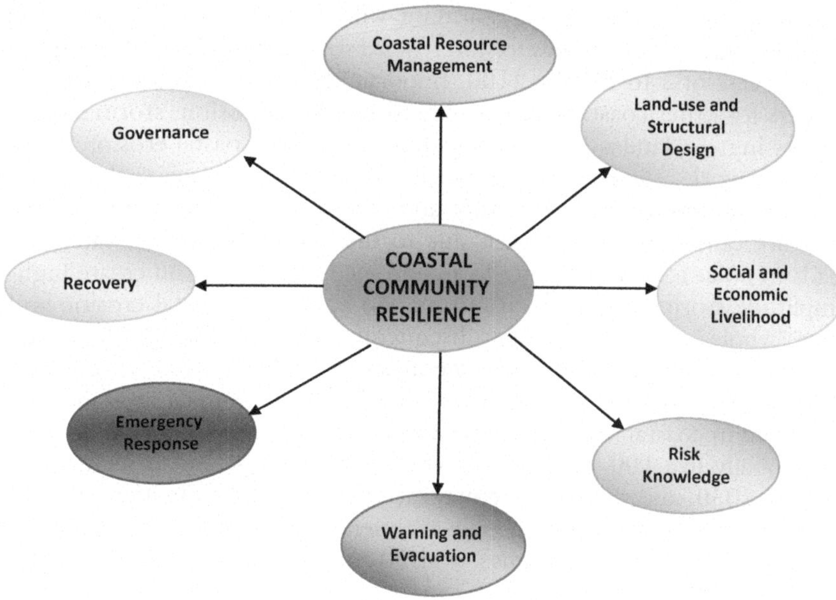

Fig. 1 Parameters of coastal community resilience (ADPC 2007)

infrastructure and urban planning that can be crucial for both mitigation and adaptation to resilience building.

Various identified dimensions, their respective parameters and variables which can be used for resilience assessment are shown in Fig. 1.

There are 13 coastal States/Union Territories encompassing 84 coastal districts which are affected by various natural hazards during different seasons in India. As per the World Meteorological Organization, the subcontinent is exposed to nearly 10% of global tropical cyclones. As is evident from past experiences, vulnerability to cyclones is experienced more in the Bay of Bengal (east coast) as compared to the Arabian Sea (west coast) as was noted after Cyclone Ockhi, which severely affected the west coast. The Inter-governmental Panel for Climate Changes'(IPCC) Third Assessment Report of 2001 clearly indicates that the factors attributed to the increase in cyclone activities in the Arabian Sea are changing weather patterns and anthropogenic emissions which are adding to the vulnerability of the west coast (IPCC 2001). Scientists such as Kelvin Wash, Jinhua Yu and Yu Qing Wang, who have researched climate change and tropical cyclones, suggest that global warming will cause the globally averaged intensity of tropical cyclones to shift toward stronger storms with an intensity increase of 2 to 11% by 2100. Importantly, tropical cyclones are harbingers for other coastal hazards such as flooding, erosion, storm surges, salinity ingress, landslides on continental slopes and coastal erosion. How does this work? The emission of greenhouse gases accelerates cyclone genesis both in frequency and intensity, and corresponding change in sea temperature causes rise in sea levels due to thermal expansion of water and leads to polar ice cap melting. As a combined effect, this can create high-amplitude storm surges which in turn can accelerate coastal erosion and alter whole shorelines and subsequently change coastal dynamics. According to IPCC's Fifth Assessment Report of 2013, extreme temperatures are expected to increase in India by 1–4°C, with maximum increase projected in coastal regions, by the 2030s. Approximately 1.2 billion people globally live 100 km from the coast. This is expected to increase by 50% by 2030, and these people may face the threat of extinction. To add to this, each year 10 million people experience coastal floods due to storm surges and land fall of cyclones, and 50 million will be at risk by 2080 (Nicholls 2004).

The east coast of India is dominated by coastal plains and is wide, with many large deltas, lagoons, mangroves, long and wide stretches of sand dunes, ridges and beaches, whereas the common geographic features along the west coast are rocks, headlands, cliffs, estuaries and bays. The National Centre for Coastal Research conducted a National Assessment of Shoreline Changes along the Indian coast in 2018 and mapped approximately 6031 kms of coastline of the total of 7500 km to analyze the temporal shoreline changes between 1990 and 2016 using nine data sets. Vulnerability mapping was done to assess erosion, stability and accretion. Out of the total, 33% of coastline is under varying degrees of coastal erosion, 29% is of natural accreting condition and the remaining 38% falls under stable state. The state-wise analysis suggested that more than 40% of erosion is noted in four states/UTs, that is, West Bengal (63%), Pondicherry (57%), Kerala (45%) and Tamil Nadu (41%) coast. Accretion exceeds 40% along the coasts of Odisha (51%) and Andhra Pradesh (42%) (National Centre for Coastal Research: Status Report 1990–2016). The west coast of India (except Kerala) is mainly in stable condition, aside from with isolated pockets of eroding coast.

Further, Indian coastal communities are increasingly at risk from many coastal hazards, as coastal habitats such as reefs, mangroves and wetlands are being destroyed by a wide range of human uses, including shoreline development, land reclamation, mining and aquaculture. Over-fishing and the use of destructive fishing practices are causing the decline of fishery resources and changes in marine ecosystems' structure and function. A major impact has been witnessed in the coastal farmland, which has deteriorated due to salinity from agriculture and farming practices, especially after disasters, as both farmland and crops are directly and severely damaged, leading to enormous economic losses. The degradation of coastal farmland and environment from chronic natural and human-induced actions threatens food security, livelihoods, and the overall economic development and well-being of coastal communities. Most of the coastal population in states such as Tamil Nadu, Andhra Pradesh, West Bengal, Maharashtra and Odisha live in relatively densely populated areas where basic services are limited, and they lack adequate capacity to plan and respond to most impending hazards, which adds to their already vulnerable status and increases disaster risk, impeding development. These conditions set the stage for more frequent and severe disasters and reduced time and capacity to recover. Thus, vulnerability to coastal hazards is increasing because of the spate of anthropogenic activities, corroborating the need for increasing community resilience.

Therefore, the aim of identifying coastal community resilience is to suggest vulnerability reduction measures accordingly for the multitude of risks to which they are exposed. For this, linkages between community developments and coastal disaster management processes and activities are needed to build coastal community resilience to coastal hazards. These linkages need to be explicit and driven by community members themselves, working in concert with national and local government programs and NGOs. Hence, for the coastal risk management arena to be able to reap the benefits of SDGs, convergence between identified and implementable coastal resilience parameters and developmental actions must be synergized. Further, it will be crucial for the implementation pathways to build coastal community resilience, and for SDGs to adopt a cross-domain approach.

Governance This aspect tries to capture details of all the essential last mile services such as schools (primary and secondary, higher educational institutes, accessibility, standard of government schools and other related aspects); hospitals (primary health care, super specialty, accessibility, quality); road networks (land transport networks and disruptions after disasters, paved roads and their accessibility during flooding and storm surges, status of interruption after calamity, drainage in roads); potable water (access, availability, supply capacity, alternative supply facility); and sanitation and solid waste disposal (access to sanitation, collection of waste, treatment, recycling, solid and debris waste management after disaster).

Coastal Resource Management The potential coastal biodiversity details are utilized, such as active bio-shields (e.g. mangrove cover, dunes, lagoons, forest cover, corals and seagrass ecosystems), resources under threat and degradation due to natural or human-induced aspects, changing coastal dynamics, mapping of critical habitats, and so on. Other details need to be addressed regarding plans and projects being implemented in such hotspots, community participation in conservation practices, and current status of coastal resource management.

Land Use and Structural Designs Here we can take note of land-use pattern (nature of houses, construction close to the coastline, status of vegetation cover, harbors and jetties, encroachment activities); violations of Coastal Regulation Zones such as construction; lack of awareness regarding impending disasters; infrastructure endangering critical habitats, and the presence of natural structural designs (e.g. mangroves, dunes, forest cover and its status); structural designs and their necessity, availability, conditions, efficiency and impacts, along with beach nourishment activities.

Social and Economic Livelihood Details regarding population (very severely exposed, severely exposed, moderately exposed, poorly exposed, very poorly exposed); education and awareness (literacy rate, population awareness about disaster, availability of awareness programs/disaster drills, access to internet, functionality of schools after disaster); health (population suffering from waterborne diseases, access to primary health facilities, access to hospitals, capacity of health facilities during disaster); community preparedness (logistics, management and materials); provision of shelter for affected people (support from NGOs, self-help groups, community participation in relief); social capital (population participating in community activities/clubs, acceptance level of community leader, communication between community and local government, level of ethnic segregation); income (population below poverty level, number of income sources per household, income derived from informal sector, percentage of households with reduced income due to a disaster); employment (percentage of unemployed labor, percentage of unemployed women, percentage of unemployed youth, percentage of employees coming from outside the area, percentage of child labor); household assets (television, mobile, motorized vehicles, non- motorized vehicles, furniture); finance and savings (availability of banking, availability of credit, accessibility to credit for poor people, saving practice of households, property insurance status); and budget and subsidy (funding for disaster risk management (DRM), budget for DRR (disaster risk reduction), availability of subsidies to rebuild houses, alternative livelihood, health care after a disaster) are assessed under this sub-heading.

Risk Knowledge It is important to comprehend the level of understanding of the concerned communities about various risks for better planning. The following require detailed attention in such assessments: knowledge and information regarding hazards (slow onset, rapid onset, natural, anthropogenic, proper understanding); individuals with special needs (populations below age 14, population above age 60, differently abled persons, pregnant women, nursing women); community structures (panchayats, NGOs, community-based organizations, youth groups, self-help groups as resources during emergencies); public awareness (risk, mitigation, preparedness, rescue, relief, community-based disaster management); trainings and drills (participation from the community, mock drills so as to assess capacity building initiatives).

Warning and Evacuation While it is impossible to plan for each disaster separately, as every emergency has a different dynamic, failure to do so can cause intense damage to assets, human life and revenue, and hence stakeholders continue to focus on the need for a systematic approach and assistance in carrying out efficient responses. This calls for better accessibility of hazard notifications, setting up of warning towers and centers, alert communications, and identification and mapping of evacuation routes, zones and shelters.

Emergency Response The aim of this 'Emergency Response' is to provide immediate assistance to maintain life. The focus is on meeting the basic needs of the people so as to contain or lessen the effects of the disastrous event, thereby preventing further loss of life and property and re-establishing normalcy. Details about quick response systems (availability, accessibility and efficiency), emergency services (medical, transportation, food, shelter and drinking water) and the role of stakeholders need to be mentioned here.

Recovery This is a long-term initiative which calls for due recognition of community participation in reconstruction and rehabilitation schemes. Details of disaster recovery plans (progress, benefits, drawbacks, effects on community); economic recovery (government agencies, co-operative organizations, property insurance); environmental recovery (beach stabilization, repairing of structural mitigation measures, bio-shield repairing); and sustainable recovery plans are to be incorporated under this sub-heading.

Assessment of the level of resilience among the coastal communities based on the above parameters and indicators can provide an insight into fulfillment of several SDGs which are interlinked with the achievement of resilience. SDG 1, regarding poverty alleviation in islands and coastal farming communities, can include well-designed social protection schemes as most poor people remain perilously close to the poverty line, experiencing shocks with little scope to manage. It also encourages better management of natural resources to strengthen their resilience by both reducing the likelihood of natural hazard events and offering resources to help cope with them. SDG 2, which aims at ending hunger and sustaining food security, could be achieved by promoting sustainable seafood and agriculture. SDG 6, which focuses on ensuring the availability and sustainable management of water and sanitation for all, can be implemented for betterment of ocean health, which can also add to the growth of coastal resources. SDG 8, to promote sustained, inclusive and sustainable economic growth and provide productive employment through the contribution of blue ocean economies, and likewise, SDGs 9 and 11, encourage building resilient infrastructure, making human settlements inclusive and safe, and fostering innovation by promoting 'Blue Revolution'.[1] These elements will thus foster better governance, land-use planning, and social and economic upliftment. Small-scale fishers constitute the largest employment category among the bottom 40% of the population by income, and hence a sustained blue economy would also address SDG 10, which advocates reducing inequality within and among countries. SDGs 12, 14 and 15 pursue sustainable consumption and production patterns of resources, products and services derived from the ocean, along with restoring forests, reversing land degradation and halting biodiversity loss, which if utilized prudently can produce efficient resource management. These three latter SDGs are directly linked to SDG 13, which calls for taking urgent action to combat climate-related hazards and their impact, especially for the coastal zones.

The priority in the current scenario is the rising density of India's population. At least 200 million people were estimated to live in the coastal floodplain in 1990, and it is likely that their number will increase to 600 million by the year 2100 (Nicholls and Mimura 1998). This both places growing demands on coastal resources and increases people's exposure to coastal hazards. India has the highest population of fishing communities globally. These communities are spread over 3600 fishing villages and more than 14.50 million people depend on fisheries for their livelihood. In order to ensure sustainable development of the sector, and with an emphasis on livelihood creation as well as resource conservation, some of the measures include establishment of a Potential Fishing Zone Advisory program, modernization and up-gradation of fishing centers, as well as banning of mechanized fishing in certain areas. Measures for a 'Blue Revolution' are being pursued, and a central plan on Integrated Development and Management of Fisheries has been formulated. The detailed Integrated National Fisheries Action Plan, 2016 (July 2016: Guidelines on Central Sector Scheme on Blue Revolution) has also been formulated and envisions connecting 15 million beneficiaries for livelihood opportunities through various interventions. Various national and sub-national legislations are in place for the management and protection of the coastal and marine environment. India has also ratified numerous international conventions related to the use of oceans and their resources, including the United Nations Convention on the Law of the Sea. An online mechanism for predicting the movement of oil spills, the Online Oil Spill Advisory System, was launched in 2015. In addition, the revised National Oil Spill Disaster Contingency Plan, 2015 reflects the important national regulations as well as the current international norms. Various levels of marine pollution are being monitored across several locations along the country's coastline through the Coastal Ocean Monitoring and Prediction System. India is also setting up a Marine Observation System along its coast to gain a better understanding of coastal processes and monitor water quality. Coastal tourism is also being promoted under the flagship Sagarmala Programme for enabling access to better livelihood opportunities.

An exploding population means rising demand for food and other utilities. Macro marine algae are considered one of the best sources of food, fodder, fertilizer, medicine and chemicals. Seaweed extract is widely used in toothpaste, ice cream, ketchup, textile printing, teeth fillings, cosmetics, tissue culture, plywood, packaging and other industries. It has also been observed that the cultivation of seaweed may be used for carbon dioxide sequestration to combat global warming and value addition to its harvest can substantially raise income. Coastal regions bear the wrath of climatic variability that hinders and creates havoc with agriculture and the livelihoods of coastal farmers. Ingressions of saline water into cultivable fields render the fields salinized and unfit for farming. Reclamation of such farmland is a slow process which can persist even after three or four monsoon seasons. Coastal farmers predominately practice the paddy-fallow cropping system and thus low-lying villages suffer immensely from inundation of paddy fields, which is a recurrent concern. In order to overcome such concerns, agro-ecosystems in cyclone- and flood-prone coastal regions can be rejuvenated through land shaping for rainwater harvesting, utilization and integration of farm enterprises. Adoption of land shaping treatment has facilitated taking up of short-duration, high-yield varieties of paddy in kharif seasons and a second crop of vegetables such as okra, chillies, tomato and brinjal during the rabi-summer season. Harvested rainwater is being used for fish and duck rearing and to provide irrigation for vegetable crops. An alternative for resource-poor farmers who own vulnerable farms is adoption of 'ail' or embankment cultivation, which involves raising an embankment all around a low-lying field with soil excavated by digging a field drainage channel all around. Vegetable crops are raised on the land embankment area, which was previously impossible in low-lying fields. The practice also helps in overcoming the salinity problem affecting crops in the rabi-summer season. Many coastal farms are reeling under drought, leading to agrarian crises. However, some farmers in coastal agricultural regions have overcome drought conditions with simple but innovative practices by reverting to traditional crops and farming methods to fight the lack of irrigation water and soil salinity. Switching to organic farming has helped improve soil quality, farm ponds have helped increase water levels in many villages, and such ponds become the source of water for cattle and vegetable farming and also for aquaculture. Floating plants and cultivating native fishes in such pond farms have helped increase income. Such solutions will become a necessity as, unlike many other anticipated consequences of climate change, global sea level rise is already taking place. The M.S. Swaminathan Research Foundation (2016: MSSRF funding on developing salt-tolerant rice) based in Chennai is now experimenting with salt tolerant rice hybrids as well as several salt-tolerant vegetables. Salt-tolerant rice varieties are now being grown in the state of Kerala, and these can grow in hyper-saline soil. Over the last 100 years, global sea levels rose by 1.0–2.5 mm/year. Disasters impact agriculture, and farmers are already experiencing multiple concerns; hence, these impacts on agrarian communities need to be woven throughout the disaster management realm. Commendable examples of resilience building and adaptation are isolated and fragmented in nature and are not uniform across India, and successful models need to be replicated.

In conclusion, coastal communities cannot be separated from their ecosystem of marine land ecology. They coexist with the natural processes of carbon sequestration in which sea vegetation, marine animals, coastal avian breeds and humans must play a role. Resilience building for coastal communities would definitely mean a developmental process which addresses the coastal disruption due to blatant defiance of coastal regulatory zones and the Disaster Management Act 2005. India is divided into approximately 90 agro-ecological climatic diversity regions and nearly 84% of its water is used for agro purposes. It thus becomes imperative to link rainfall patterns with cultivation practices and patterns for sustainable agriculture which will invariably lead to promotion of reverse migration. There is also an emerging need to enhance the earning capacity of the people living in the coastal areas. Sea farming can be implemented as a new model for sustainable livelihood in coastal India, and is being encouraged through large-scale seaweed cultivation as an alternate and additional means of livelihood. Present estimates of future sea level rise range from 20 to 86 cm by the year 2100. Therefore, the adaptive capacities of inhabitants need to be strengthened by creating effective mitigation policies and building resilience. In addition, an integrated holistic approach toward farming as an occupation rather than providing ad hoc solutions must be ensured. Along with addressing social vulnerabilities, the economic resilience of communities needs to be strengthened. This can be achieved by using micro-insurance as a tool to protect low-income populations against specific perils, empowering self-help groups and providing adequate connections between micro-mandis and larger markets so that perishable goods do not end up being wasted and can benefit many small and marginal farmers. Second, governance must advocate enhancing the decision support system for such communities. Here, the last mile connectivity should not be limited to major events but must focus on providing information on day-to-day activities. For example, agro-advisories should be linked to weather updates and provide interpretations for better farming and cropping patterns. Third, to avoid stress on the primary occupations held by coastal communities, diversification of income opportunities must be explored through provision of skill-based education such as tapping into coastal tourism as a potential option. Replicating good examples such as bio-energy villages,[2] a system which is practiced across many villages globally, can ensure holistic development of such communities.

Notes

1. This refers to the period of intense growth in the world-wide aquaculture industry from the mid-1960s to the present. World-wide aquaculture production has now reached 50 million tons, up from 0.2 million in 1950.
2. This is a regionally oriented concept for the use of renewable energy sources in rural areas. The system uses biomass from local agriculture and forestry in a biogas power plant to meet the complete energy requirements of a village, such as electricity and district heating, thereby promoting self-sustainability.

References

Achieving the 2030 Agenda and the Sustainable Development Goals for Oceans and Coasts in the Western Indian Ocean Paper for the Science-Policy Workshop of the 9th Conference of Parties to the Nairobi Convention, 9–11 July 2018, Durban South Africa; Yvonne Waweru.

ADPC (Asian Disaster Management Centre). (2007). Emerging Risks and Approaches to Reduce Vulnerability in Urban Built in Environment. *Asian Disaster Management News*. Available at http://drr.upeace.org/english/documents/References/Topic%203-DRR%20Processes%20and%20Hyogo%20Framework%20of%20Action//ADPC%202007%20Reducing%20Vulnerability%20of%20Urban.pdf. Accessed 5 Nov 2019.

Climate Change Threatens 55 Million in India's Coastal Areas Web: https://thewire.in/environment/climate-change-threatens-55-million-in-indias-coastal-areas-report-warns. Accessed Oct 2018.

Coastal Zone Management in India – Present Status and Future Needs; Shailesh Naik: June 2017, pp. 174–183. Accessed Sept 2018 Web: https://www.tandfonline.com/doi/full/10.1080/10095020.2017.1333715

Guidelines: Central Sector Scheme on Blue Revolution: Integrated Development and Management of Fisheries; Department of Animal Husbandry, Dairying & Fisheries Ministry of Agriculture and Farmers Welfare Government of India: National Fisheries Development Board; July 2016. Web: http://nfdb.gov.in/PDF/GUIDELINES/Guidelines%20CSS%20on%20Blue%20Revolution_Integrated%20Development%20&%20Management%20of%20Fisheries_June%202016.pdf. Accessed Sept 2018.

Hoozemans FMJ, Marchand M, Pennekamp HA (1993). A global vulnerability analysis, vulnerability assessments for population, coastal wetlands and rice production on a global scale, 2nd edn. Delft Hydraulics and Rijkswater-staat, Delft.

How Resilient Is Your Coastal Community? A Guide for Evaluating Coastal Community Resilience to Tsunamis and Other Hazards; U.S. Indian Ocean Tsunami Warning System Programme, Printed in Bangkok, Thailand; 2007, pp. 10–164.

IPCC (2001). Third Assessment Report: Climate Change. Available at, https://www.ipcc.ch/report/ar3/wg1/. Accessed Sept 2019.

M.S. Swaminathan Research Foundation; Centre for Research on Sustainable Agriculture and rural Development: 'Salt tolerant Rice'; April 2016 Web: https://www.mssrf.org/mssrfoldsite/?q=tags/salt-tolerant-rice. Accessed Sept 2018.

National Assessment of Shoreline Changes along Indian Coast; R.S. Kankara, M.V. Ramana and M. Rajeevan, Status Report for 26 Years – 1990–2016: Ministry of Earth Sciences and National Centre for Coastal Research, Chennai; July 2018.

Nicholls, Robert J., (2004). Coastal flooding and wetland loss in the 21st century: changes under the SRES climate and socio-economic scenarios, Global Environmental Change 14 (2004) 69–86.

Nicholls, R.J., & Mimura, N. (1998). Regional issues raised by sea-level rise and their policy implications. *Climate Research*, 11, 5–18.

Voluntary National Review Report on the Implementation of Sustainable Development Goals: United Nations; High Level Political Forum 2017: Report to Be Presented to The High-Level Political Forum on Sustainable Development, New York, July 2017.

Tamil Nadu Farmers Fight Drought with Organic Farming; Sharada Balasubramanian: Web: https://www.aljazeera.com/indepth/inpictures/2017/04/tamil-nadu-farmers-fight-drought-organic-farming-170424091238696.html. Accessed Oct 2018.

Sushma Guleria is an assistant professor with the Environmental Disaster Risk Management Division National Institute of Disaster Management, under the Ministry of Home Affairs at New Delhi. She received a PhD with a thesis entitled

'Integrated Coastal Zone Management Plan for Tsunami Affected Coastal Areas in Cuddalore, Nagapattinam and Kanyakumari Districts, Tamil Nadu, Southern India' from Suganthi Devadason Marine Research Institute, Tuticorin, Tamil Nadu. She has worked on a World Bank project on School Environmental Education in India with the Ministry of Environment, Forest and Climate Change and has over 12 years of experience in the Disaster Management realm with involvement in Training and Capacity Building programs, Module Development, documentation of disaster events, developing policy briefs and Course Curriculum. She has more than 21 publications to her credit, including books, and international and national papers. In her profile, she looks after the Centre for Climate Resilience and Environment and Centre for Water and Land Disaster Risk Reduction (DRR) with a focus on eco-DRR, climate- and environment-related disasters, policies and tools, community resilience, livelihood security and natural resource management. She was the recipient of the World Bank's Young Researchers Grant Award for Disaster Risk Reduction for the year 2005–06 from India.

Disasters and Climate Change Adaptability at Odisha Coast

Niranjan Sahoo and Maheswar Satpathy

Capturing a Recurrence and Unpredictability of Disasters at Odisha Coast

Historically, Odisha has remained one of the most vulnerable Indian states to experience natural hazards, particularly cyclones due to its sub-tropical littoral location (OSDMA 2016). It is no mystery that third-world countries like India have low adaptational capacity to combat the inimical effect of climate change. Climate change has emerged as a major multiplier of disaster losses worldwide. These environmental disasters are those extreme events induced by nature that exceed the tolerable magnitude and make human adjustment very difficult, resulting in colossal loss of property, human and animal lives, destruction of settlement and environment. The coastal resources are suffering from a sustained net decline, largely related to coastal squeeze of intertidal habitats (Carpenter and Pye 1996). As

N. Sahoo (✉)
Department of Geography, Utkal University, Bhubaneswar, India

M. Satpathy
University College London (UCL), London, UK

A. Singh et al. (eds.), *Development in Coastal Zones and Disaster Management*, Disaster Research and Management Series on the Global South, https://doi.org/10.1007/978-981-15-4294-7_12

Odisha has the longest coastal belt in the Indian sub-continent, that is, 499 kilometres, vastly covering the Bay of Bengal region, it is necessary to adopt different types of remedial measures to overcome the casualties as well as frequency of disasters in coastal Odisha.

Bay of Bengal bears enormous climatic importance. It is important to note that, on an average, Bay of Bengal is the only sea in the world which records super cyclones every two to three years. Patel (2016) has delineated climate change-related issues in his paper 'Climate Change and Climate-Induced Disasters in Odisha, Eastern India: Impacts, Adaptation and Future Policy Implications'. He has provided a historical analysis of climate change in Odisha and its impact on health and poverty in the coastal region. The findings of the study highlight that there is scarcity of research which can focus on different climate induced calamity issues. He highlighted the importance of different adaptive measures, such as, strong political will, strengthening infrastructures, public private partnership, community involvement and better planning for risk management, to look at actions that address climate change and deliver benefits for growth and development. He argues to demand appropriate research and policy advocacy to find better solutions and development strategies in the state.

The Odisha coast is affected to a greater degree due to the frequency and intensity of turbulences and disasters. This affects the region's ability to bounce back and how it responds to critical threats of facing an imminent calamity. Olmos (2001) pointed out that the propensity of a system to adapt to impacts of climate change is known as its adaptive capacity and is influenced by certain characteristics known as determinants of adaptation, which may be contextually embedded as well as determined by the nature of that calamity. Similarly, Holling (1973) argued about the concept of adaptive governance to encounter climate change. Thomalla and Schmuck (2004) examined the loss of property, preparedness and current practices of disaster management since Super Cyclone 1999. More work on adaptational capability finds its roots in the work on socio-ecological systems and adaptive management as proposed by Olsson et al. (2004). Their primary focus was dynamic multilevel and polycentric management. In the same way, Folk (2006) suggested ways to tackle environmental changes through climate change and ways to formulate adaptive measures against this backdrop.

The author collected data using a series of in-depth interviews with participants who have been experiencing frequent cyclonic outbursts and have seen a transition of the state since Super Cyclone 1999. A thematic content analysis was applied using secondary data such as newspaper articles and different research pieces published in scholarly journals. Interviews consisted of questions related to their responses and response-preparedness as it evolved during and after each disaster. Similarly, data on their anguish, suffering and losses, which were a direct result of a disaster, were collected. Many participants. who engaged in a didactic brainstorming session, were interviewed and observed during the process. A semi-structured pre-post-questionnaire and group discussion (participatory design method and brainstorming) followed. Qualitative Data was analysed using content analysis and generic thematic methods.

Climate Change and Disaster in Coastal Odisha

In an age of climate change, we have witnessed a rise in the mean temperature of the globe, polar ice cap melting and rising sea level, exposing the coastal belt inhabitants to greater vulnerability and unpredictability. Climatologists and different researchers in this field claim that world is at risk because of climate change, as it is abundantly clear that it is causally linked with different forms of disaster. However, climate change and management of resultant disasters has not occupied mainstream interest in a developing country like India despite the third assessment report of the IPCC (2014) that developing countries are more prone to impacts of climate change. Owing to the geographical location of Odisha, the adverse impact of climate change would grow likewise. Meanwhile, the state has experienced Super Cyclone in 1999, Phailin in 2013, Hudhud in 2014 and Titli in 2018. We can narrate that severity and frequency of disasters continue to grow at a rapid rate under the influence of global climate change.

Catastrophic Cyclones at Coastal Odisha

Since 1999, when Odisha experienced the so-called super cyclone, it has continued to experience several small- and medium-range cyclones, for example, Phailin in 2013, Hudhud in 2014 and Titli in 2018. We can hypothesize that the severity and frequency of cyclones are growing at an unprecedented rate. Any tropical cyclone that develops in the north of the Indian ocean between 100°E and 45°E is monitored by the Indian Meteorological Department[1] (IMD, New Delhi) A look into the three cyclonic disasters which occurred in Odisha may reveal the state preparedness and challenges.

Titli is one of the recent cyclones which made landfall near Gopalpur in Ganjam district of Odisha on the early morning of 12 October 2018. Its wind speed reached up to 135 kmph and at times touched about 165 kmph. The coastal areas and adjacent interior regions of Odisha received heavy torrential downpour. The death toll rose to 57 people and claimed a large number of livestock and other animals. The high-speed winds had caused damage to the socio-economic condition of the state. It washed away *kuchha* or mud houses, trees and electricity poles. The state government, along with social activists and NGOs, helped in an evacuation of over three lakh people residing in coastal Odisha as a preventive measure. This has been one of the largest preparedness measures undertaken by any state till today.

The Odisha coastline is not strong enough to withstand the severe cyclonic storms. Superstorms like Phailin, Hudhud and the infamous 1999 Odisha cyclone have lashed the region in the past, claiming thousands of lives and uprooting countless homes. Around 300000 people were evacuated and 879 multipurpose cyclone and flood shelters were kept in readiness for sheltering the rescued people.

In 2014, Hudhud originated in the Andaman sea. It was a category 4 tropical cyclone (extremely severe) that had a landfall near Visakhapatnam of Andhra Pradesh with a wind speed of 185 kmph. The districts placed under high alert by State Disaster Management Authority (SDRF), Odisha government are: Balasore, Kendrapara, Bhadrak, Jagatsinghpur, Puri, Ganjam, Mayurbhanj, Jajpur, Cuttack, Khurdha, Nayagarh, Gajapati, Dhenkanal, Keonjhar, Malkangiri and Koraput. Southern Odisha experienced heavy rainfall and wind speeds reached up to 90 kmph, without much losses to vegetation or livestock. The government had evacuated 67,752 people to safe places and cyclone centres in the southern districts before the storm reached Odisha. The very severe cyclonic storm Hudhud impacted 247,557 hectares of agricultural land in Odisha of which 40,484.5 hectares have sustained crop loss of over 50%. A preliminary damage assessment report by the state agriculture department has pegged the crop loss at Rs 23.77 crores. The agriculture department had sought an assistance of Rs 4500 per ha for crops damaged in rain-fed area, Rs 9000 per ha for irrigated area and Rs. 12,000 per hectare for perennial area. The damage to electrical infrastructure alone by Hudhud in 2014 had been assessed at Rs. 104.35 crores.

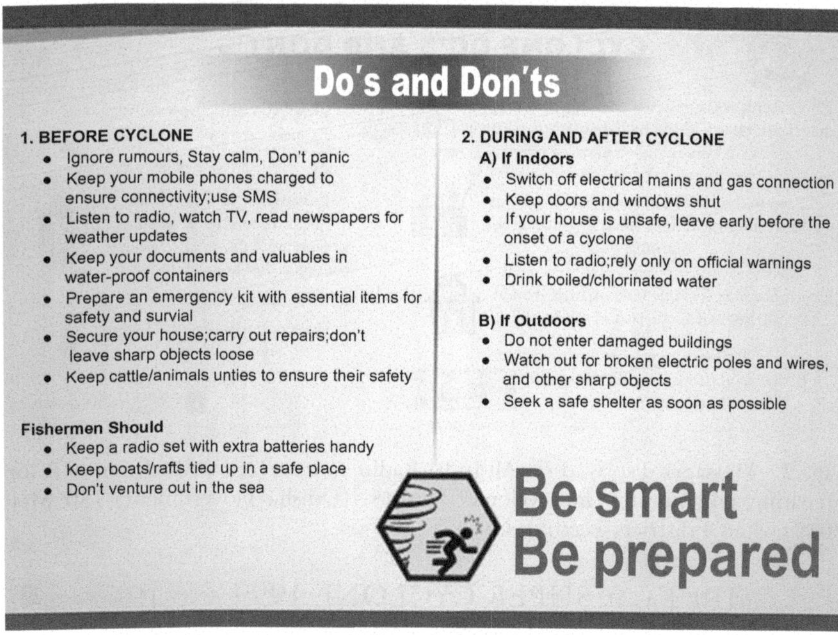

Fig. 1 Preparing communities for cyclonic disasters. (Odisha Government)

Phailin in 2013 was the most intense cyclonic storm which made landfall in Gopalpur. The cyclone triggered India's biggest evacuation drive in 23 years as more than 550,000 people were evacuated from the coastal tracts to safer regions. Trees and electricity poles were uprooted due to the windy conditions even in interior regions. Phailin also caused immense flooding and damaged crops in over 500,000 hectares of agricultural land in the state (Fig. 1 and 2).

The concern which emerges here is that of a committed need for preparedness because the cyclones are severe and even the most efficient government cannot rush at the last moment; a major share of preparedness ought to precede the disaster and almost in permanence. The Odisha government could handle the disasters with much relative efficiency due to pre-existing shelters, storehouses, department-based rapid action teams and an aware community which does not waste time in alerts and explanations but just follows administration in trust and faith to a safer region.

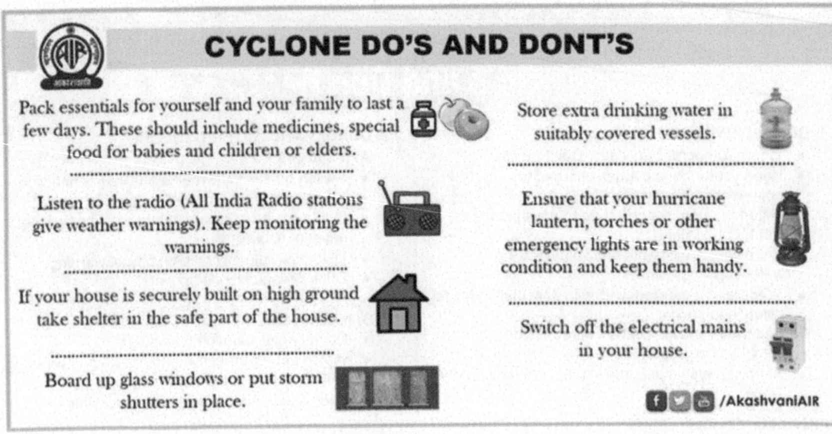

Fig. 2 Messages displayed on All India Radio and tweeted @AkashwaniAIR for preparing communities for cyclonic disasters. (Odisha Government) (30 May 2017) when 4 districts were put on alert

The Fatal SUPER CYCLONE 1999 and the Lessons Learnt

Odisha had not always been a prepared state prior to the Super Cyclone 1999. The 1999 cyclone, which changed disaster governance in Odisha, was the strongest and most destructive cyclone ever recorded in the history of Odisha. It reached super cyclonic storm intensity with wind speeds reaching up to 260 kmph a day before it made its landfall. The region received torrential rain and widespread flooding with a surge carrying coastal debris to inundate entire towns and villages. Most of the wintertime harvest was destroyed. The government estimated the official death toll as 9887. The unofficial death toll was considerably higher. Thousands of people and livestock lost their lives in the rising storm surge. Over 8000 deaths occurred in Jagatsinghpur alone. Union Government declared it as a national disaster as 12 districts in Odisha namely Balasore, Bhadrak, Cuttack, Dhenkanal, Jagatsinghpur, Jajpur, Keonjhar, Kendrapada, Khurda, Puri, Mayurbhanj, and Nayagarh were worse affected. In the aftermath, many epidemics and diseases spread through the state as well. The Odisha government was severely criticized for its apparent lack of preparedness and inadequate response to the cyclone (Thomalla and Schmuck 2004). This criticism led to the step 'one' of mapping hazard zones and vulnerabilities (Fig. 3).

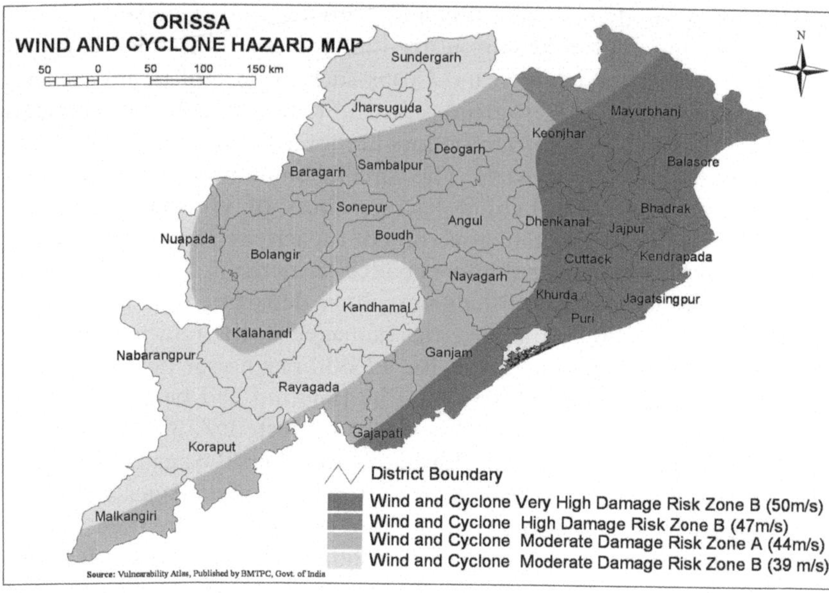

Fig. 3 Wind and cyclone hazard map of Orissa. (Source: Vulnerability Atlas, Published by BMTPC, Government of India)

As cyclone is associated with torrential rainfall, the aftermath of cyclone is marked by torrential rainfall spell across the coastal region and interior parts of the state. The socio-economic condition of people, their livelihood and loss of standing crops, houses and environmental degradation makes them highly vulnerable.

Impact on Coastal Livelihood

A disaster induced by the cyclone results in heavy rainfall, destroys harvests, washes away roads, houses and also sea resources. It also causes inundation of an area with salt water for months, thereby making the landmass infertile for generations.

The loss in agriculture takes place as a two-way process: strong wind and flood wipe out standing crops in the kharif season. The cyclonic depression over Bay of Bengal has been triggering heavy rains across coastal Odisha. The districts which often suffer from crop damage due to cyclones are primarily Balosore, Bhadrak, Cuttack, Puri, Jagatsingpur and Ganjam. The inundation of salty water into the coastal low land turned fertile land barren for many years. Crops such as that of paddy, several vegetables, sugarcane, pulses, groundnut and cotton mostly get damaged in coastal Odisha. This type of crop loss has forced people to emigrate to the other states in search of work. This forceful emigration of workers and farmers is another cause behind backwardness of Odisha's economy. Moreover, the lack of the state government capacity to safeguard or compensate through a matching support created farmers unrest and continual suicides in coastal Odisha. Odisha's agriculture department had sought an assistance of Rs. 4500 per ha for crops damaged in the rain-fed area, Rs. 9000 per ha for irrigated area and Rs. 12,000 per ha for perennial area. The damage to electrical infrastructure by Hudhud had been assessed at Rs. 104.35 crores. The cyclonic storm had affected 773,691 electrical consumers, mainly in southern Odisha districts like Rayagada, Koraput, Gajapati, Nabarangpur and Malkangiri. A total cost of damages and losses taken together may run into several millions of rupees. Since then, several assistance procedures have been lent by neighbouring countries, developmental aid organizations and bilateral agencies. The seed-cum-grain bank programme envisaged the revitalization of existing women's Self Help Groups (SHGs) to enable the preservation and growth of local rice with inputs like seeds, seed bins and skill upgradation in one of the worst-affected coastal block of Odisha (Swain 2002). A cumulative effect of disaster on environment is wiping out of coastal biodiversity and with it rare ecosystems preserved in water bodies such as the lake Chilika is fatally encountered in Odisha.

Coastal Odisha is famous for fish-culture. Generally, fishermen are not willing to ferry inside the ocean when IMD issues warnings. Consequently, fish business gets stuck for many days, throwing several daily wage labourers into unemployment. Cyclonic disasters for the lake Chilika become a huge setback for Chilika-specific fish-culture. The flood also prevents the fish-culture due to the flood water remaining high for over three months. Most fisher folk stay unemployed for months and sink into debt, thus out-migrating as cheap labour for big cities.

Human Adaption to Disaster in Coastal Odisha

Despite the losses and damages, everything has not been as depressing as it appears to be. Disasters and development have learnt to co-exist in Odisha. Since 2000, Odisha has seen an unprecedented economic growth of an average 8%, which is much higher than the national average growth (World Bank 2016). Odisha has demonstrated resilience and continuous growth and has learnt to bounce back each time from one disaster to another. Odisha has reduced the gap with the other states by achieving a 20% reduction in poverty (World Bank 2016). The microfinance assist was in part a response to the 1974 famine (Yunus 1998). The Government of India has come up with Food Security Bill which was synchronized with social welfare schemes, policies and programmes. However, much effort is required to arrange work for people who lost their livelihood under functional national schemes like MGNREGS (Mahatma Gandhi National Rural Employment Guarantee Scheme), SGSY (Swarna Jayanti Gram Swarozgar Yojana, self-employment schemes for local communities and self-help groups) and NRLM (National Rural Livelihood Mission).

The cyclone and aftermath of flood also brings many diseases such as diarrhoea, cholera and malaria. The cost of mitigating health impacts of disasters has rarely been included in disaster management plans. There is need to adopt a more holistic plan for the recreation and development of social security systems to stop transient diseases and hunger caused by floods, droughts and other natural disaster (Dreze and Sen 1990).

Adaptation to climate change is the process through which people prevent negative impact of climate change on their wealth and well-being (Olmos 2001). The importance of adaptive governance is widely recognized and focused in the disaster risk reduction (DRR) and Ecosystem Resources Management (Chaffin et al. 2014; Dietz et al. 2003; Foerster 2011; Munaretto et al. 2014; Singh 2017). Both adaptation and mitigation have been key strategies in the domain of disaster risk reduction and climate change strategies. The government of India has been implementing many polices in support of disaster management since 1999 and the geographical location of Odisha and climate change contributes to many positive outcomes of frequent disasters (Bhatta 1997). Odisha significantly lags behind in the adoption and implementation of latest technologies to prevent the occurrence of disaster as adaption methods. For example, the structure of Panchayati Raj body would be a key factor in disaster management in the rural areas. In a country like India, Panchayati Raj body of rural areas would play a crucial role during disaster (Singh 2008). Poor people are more vulnerable to disaster as they have least access to resources (IPCC 2014).

Adaptation to the unprecedented and unpredictable journeys of disasters need to be understood by making science adapt to local indigenous wisdom as well. The warning issued for cyclone Titli on behalf of the country's top weather monitoring station IMD could not help as the cyclone modified its journey during the final stage of its landfall. It also failed the mission of zero casualty of state government. Hence positive open-mindedness is required to overcome the devastation caused by disaster. An evaluation of possible damages in advance ought to be conducted for saving more lives and their livelihood. Peoples' adaptive capacity should enable people to resume work and regain vitality after all the odds. Unless these are enabled and instituted, one cannot draw a linkage to address a problem holistically and effectively. In order to achieve a broader type of adaptation, social political and environmental capacity building need to be integrated in the disaster management programmes (Ray-Bennett 2009). However, a clear rehabilitation plan with geographical focus and prioritization of intervention has to be prepared along three measurable indicators (a) infrastructure, (b) livelihood and (c) habitat. A rapid adoption of new technologies is evident and India's first automatic coastal warning for disasters has come up in Odisha (India Today 2017).

Recommended Measures and Action

Prudent and judicious adaption measures are required during the disaster and after the disaster starting from building awareness, sensitization, disaster counselling and support to physical rescuing and evacuation. Adaption measures of the nature of structural, human and institutional are needed to prevent damages and losses during disasters.

Pre-Disaster

There are some recommended best practices which create response preparedness among the public officials as well as public. District-wise as well as block-wise cyclone shelters should be constructed in disaster-prone areas of coastal Odisha well in advance. People should be made aware of the need and utility of an insurance policy. In order to create public awareness, the Government of Orissa in collaboration with the United Nations organized a state level workshop on 11 December 1999. If people lose their loved ones to the disaster, they should be accorded psychosocial and emotional and moral support besides financial and material support. Reduction and mitigation of the impact of disasters is a long process and should emphasize preparedness. (Swain et al. 2007).

Strategic mock drill should be carried out throughout the year by which people would be acquainted with strategies to tackle any possible disaster in future. Disaster warnings should be circulated by the state government using rapid media and communication technologies. Special warning should be issued to the farmers by which they can secure their standing crops. The state government should store adequate dry food, medicine, fresh water and should supply these items well in advance. Preparedness during a no-disaster period is of enormous importance which includes installation and maintenance of early warning systems, strengthening an integrated coastal zone management and awareness-generation programmes, capacity-building of all stakeholders and mangroves plantation in the coastal periphery to help mitigate the menace of climate change and disasters.

During-Disaster

The state government should keep its team together during the occurrence of disasters in coastal Odisha so that they can actively participate in the protection of coastal communities, fisherman and poor people and in their rescue and evacuation. Delivery of food, medicine and drinking water should be streamlined by balancing its demand and supply. Special care should be given to pregnant women, children and elderly as well as physically challenged. Widespread use of social media and hourly news update regarding the disaster and any loss should be made, and accessible contact with an office bearer should be maintained. Disaster management authority should take care of the weaker sections and most needy people (Thomalla and Schmuck 2004). There should not be any information gap, as its been observed that wider information gap increases one's vulnerability to be affected by a disaster. Outreach activities and investments should be encouraged to conduct research to improve the knowledge, programme and policy on climate change and disasters. During disasters, it has been observed that the ethical norms necessary to maintain the dignity of the individual and the community, which are usually compromised, should not be.

Post-Disaster

Post-disaster management includes reconstruction, assessment of damages and losses, economic rebuilding and confidence building. Mismanagement of disaster indicates that adaptive governance suffers from some shortcomings. Sometimes authorities ignore ground realities which tend to increase vulnerability of people in that area (Shinn 2016). A major problem of post-disaster management is that there is a lack of community participation. Therefore, adaptive governance framework should align with diverse socio-economic and ecological stakeholders for seeking participation (Olsson et al. 2004). Provision of alternative means of transportation and energy as well as psychological stress and shock counselling and treatment should be arranged for. Identification of victims for compensation and relief should be handed over to third parties. The insurance and other support mechanisms may run into huge payments and therefore a technically skilled group of experts should be given this task of identification. In the Phailin cyclone, crop insurances took a hit of Rs. 2000 crores owing to the destruction of crop-fields, especially in the coastal areas of Odisha (Business Standard 2014). The government of India supplied 66,668 vegetable seeds mini kits to the farmers with a total expenditure of Rs. 1 crore. Free ploughing was also provided for Rabi crop over 4451 hectares of land and in the year 2000 Kharif season, Rs. 33.10 lakh were sanctioned for Kharif crops. The period post disaster therefore becomes a period of intense politicization of the roles and responsibilities of the state and central governments.

Conclusion

It is undeniably true that climate change is triggered by human action of living an unsustainable lifestyle. The coast is particularly prone to multiple hazards like erosion, flooding, tsunamis, storm surges and sea level rise. The extent to which these occur is likely to increase under a scenario of climate change. The effects of climate change on health are likely to be predominately negative and impact most heavily on low-income countries where capacity to adapt is weakest, but also on the most vulnerable groups in developed countries. Odisha presents a case study of how a small area over the global platform can learn from disasters and make policies inclusive and effective by focusing upon preparedness.

Note

1. As per IMD, the lowest official classification used in the North Indian Ocean is a Depression, which has 3-minute sustained wind speeds of between 17 and 27 kn (20–31 mph; 31–49 km/h). Further to this is Deep Depression, which has winds between 28 and 33 kn (32–38 mph; 50–61 km/h). Gradually, it can develop into a gale-force wind speeds of between 34 and 47 kn (39–54 mph; 62–88 km/h). A Severe Cyclonic Storms have storm force wind speeds of between 48 and 63 kn (55–72 mph; 89–117 km/h), while Very Severe Cyclonic Storms have hurricane-force winds of 64–89 kn (73–102 mph; 118–166 km/h). Extremely Severe Cyclonic Storms have hurricane-force winds of 90–119 kn (166–221 km/h, 104–137 mph) and the highest classification used in the North Indian Ocean is a Super Cyclonic Storm, which have hurricane-force winds of above 120 kn (138 mph; 222 km/h).

References

Bhatta, B. B. (1997). *The Natural Calamities of Orissa in the 19th Century*. New Delhi: Commonwealth Publisher.

Business Standard. Cyclone Hudhud: Insurance Claims to Cross Rs 4,000 cr 15 October 2014

Carpenter, K., & Pye, K. (1996). *Salt Marsh Changes in England and Wales – Its History and Causes* (Environment Agency R&D Technical Report W12). Marlow: HR Wallingford & Foundation for Water Resources.

Chaffin, B., Gosnell, H., & Cosens, A. (2014). A Decade of Adaptive Governance Scholarship: Synthesis and Future Directions. *Ecology and Society, 19*(3), 56.

Dietz, T., Ostrom, E., & Stern, P. C. (2003). The Struggle to Govern the Commons. *Science, 302*(5652), 1907–1912.

Dreze, J., & Sen, A. (1990). Introduction. In J. Dreze & A. Sen (Eds.), *The Political Economy of Hunger*.

Drèze, J., Sen, A. & Hussain, A. (1995). The Political Economy of Hunger: Selected Essays. WIDER Studies in Development Economics. Oxford: Oxford University Press.

Foerster, A. (2011). Developing Purposeful and Adaptive Institutions for Effective Environmental Water Governance. *Water Resources Management, 25*, 4005–4018.

Folke, C. (2006). Resilience: The Emergence of a Perspective for social–ecological systems analyses, Global Environmental Change, 16(3), 253–267.

Holling, C. S. (1973). Resilience and Stability of Ecological Systems. *Annual Review of Ecology and Systematics, 4*(1), 1–23.

India today. (2017, June 2). New Delhi. Accessed Online on 28th September, 2018.

IPCC. (2014). Climate Change 2014: Impacts, Adaptation, and Vulnerability. Part A: Global and Sectoral Aspects. In *Contribution of Working Group II to the Fifth Assessment Report of the Intergovernmental Panel on Climate Change* (pp. 709–754). Cambridge/New York: Cambridge University Press.

Munaretto, S., Siciliano, G., & Turvani, M. (2014). Integrating Adaptive Governance and Participatory Multicriteria Methods: A Framework for Climate Adaptation Governance. *Ecology and Society, 19*(20), 74.

Odisha State Disaster Management Authority (OSDMA). (2016). *Orissa Vulnerability to Different Disasters.* Available from http://www.osdma.org/ViewDetails.aspx?vchglinkid=GL001&vchplinkid=PL003

Olmos, S. (2001). *Vulnerability and Adaptation to Climate Change: Concepts, Issues, Assessment Methods.* Climate Change Knowledge Network. Retrieved from http://www.iisd.org/cckn/pdf/va_foundation_final.pdf

Olsson, P., Folke, C., & Berkes, F. (2004). Adaptive Comanagement for Building Resilience in Social-Ecological Systems. *Environmental Management, 34*(1), 75–90.

Patel, S. K. (2016). Climate Change and Climate-Induced Disasters in Odisha, Eastern India: Impacts, Adaptation and Future Policy Implications. *International Journal of Humanities and Social Science Invention, 5*(8), 60–63.

Ray-Bennett, N. S. (2009). Multiple Disasters and Policy Responses in Pre- and Post-Independence Orissa, India. *Disasters, 33*(2), 274–290. https://doi.org/10.1111/j.1467-7717.2008.01074.x.

Shinn, J. E. (2016). Adaptive Environmental Governance of Changing Social-Ecological Systems: Empirical Insights from the Okavango Delta, Botswana. *Global Environmental Change, 40*, 50–59.

Singh, S. K. (2008). Role of Panchyat Raj Institution in Disaster Management- A Suggestive Framework.

Singh, A. (Ed.). (2017). *Disaster Law; Emerging Thresholds.* New Delhi: Routledge.

Swain, M. (2002). *Impact of Super Cyclone on Life and Livelihood of Women, An Analytical Study of Two Coastal Districts of Orissa.* Bhubaneswar: NCDS.

Swain, M., et al. (2007). *Gender Perspective in Disaster Management* (pp. 3–7). New Delhi: Serial Publication.

Thomalla, F., & Schmuck, H. (2004). 'We All Knew That a Cyclone Was Coming': Disaster Preparedness and the Cyclone of 1999 in Orissa, India. *Disasters, 28*(4), 373–387.

World Bank. (2016). *Odisha State Overview.* Available from http://web.worldbank.org/archive/website01291/WEB/0__C-349.HTM

Yunus, M. (1998). Poverty Alleviation: Is Economics Any Help? Lessons from the Grameen Bank Experience. *Journal of International Affairs, 52*(1), 47–65.

Mr. Niranjan Sahoo has graduated from the PG Department of Geography, Utkal University, Bhubaneswar, India. He has been researching on Sustainability, Climate Change and Behavioral Economics at the TREES Lab for the past 2 years. He has published in reputed journals and is a recent recipient of Harvard School of Public Health's Training Course in Environment and Climate Change, in collaboration with PHFI, Gurgaon.

Dr. Maheswar Satpathy is a research scientist at the highly prestigious University College London (UCL), London, UK. He is an emerging scholar in Global Health, Global Mental Health & Critical Development Studies. Besides this, he also serves as an executive director of Udyam Foundation-Global Association for Sustainable Development. He works in cross-cutting areas of sustainability and Global Health Policy in LMICs, and is a recipient of several prestigious fellowships including Open Society Fellowship, British Academy Grant Fellowship, AusAID's most prestigious Australian Leadership Awards, International Union of Psychological Sciences' Emerging Psychologists of the World Award to name a few. As the founder of TREES Lab, he serves as lead PI of the Pan-Indian Study on 'KAP Mapping on SDGs- A Youth Civic Engagement Perspective', which is now under operation with a sample size of 30,000+ from all over India. He has an H-Index of 25, and Citation Index of 11,000+ (Scopus Clarivate Analytics and ResearchGate), and has published more than 120 articles in high-impact journals like The Lancet, NEJM, JAMA, BMJ, BMC. He serves on the editorial and reviewers board of more than twenty reputed journals.

Women in 2018 Kerala Floods: A Sociological Narrative

Nisha Jose and Sony Kunjappan

Introduction

Every disaster has a story, a story of misfortune, misery, suffering, loss and lessons which are learnt through encounters with the impact of disasters on life, property and loved ones. Disasters always impact the poor and the vulnerable more than the privileged rich. The Nepal Earthquake of 2015 affected 8 million people, almost a third of its population. Cyclone Pam affected more than half the population of Vanuatu, Ebola outbreak in Western Africa killed more than 11,000 people. In all these disasters it is well substantiated how the poor and the vulnerable have borne the price (Nayagan 2018).

India too has had her share of disasters on account of its unique geoclimatic conditions. Floods, droughts, cyclones, earthquakes and landslides have been a recurrent phenomenon. About 60% of the landmass is prone to earthquakes of various intensities, 40 million hectares is prone to floods,

N. Jose (✉)
Social Enabler, Kottayam, India

S. Kunjappan
Central University of Gujarat, Gandhinagar, India

A. Singh et al. (eds.), *Development in Coastal Zones and Disaster Management*, Disaster Research and Management Series on the Global South, https://doi.org/10.1007/978-981-15-4294-7_13

approximately 8% of the total area is prone to cyclones and 68% of the area is susceptible to drought (Ministry of Home Affairs 2004). About 5700 kilometres of India's coastline is exposed to cyclones of severe intensity and magnitude. Kerala shares 10.18% of this sea coast which stretches north-south along a coastline of 580 km with a varying width of 35 to 120 km (Kerala 2018). Kerala has suffered three major disasters in the past, the Great Floods of 1924, Tsunami in 2013 and Ockhi in 2018. The recent Kerala Flood in August 2018 is yet another disaster that has affected the State badly.

The present study was undertaken with the following objectives:

- To critically analyse the strengths and weaknesses of women—during and after the flood.
- To identify the issues faced specifically by women in all three stages: *Rescue, Relief* and *Recovery* stages during and after the flood.
- To understand the concept of *freedom and capability* in the context of women, specifically during the 2018 floods.
- To propose policy perspectives for making women resilient towards disasters.

Disasters that have struck any country or state have tremendously affected human lives, be it women, men and children, experiencing their impacts differently. Every disaster affects women quite adversely and their already crumbling developmental indicators sink further to create a developmental trap. **The Gender Inequality Index (GII) issued by UNDP every year since 2010 has placed India in the position of 125 in a list of 159 nations (Jahan 2016; Human Development Report 2016). The GII captures three core areas of gender inequality, (1) reproductive health, (2) empowerment (based on parliamentary seats occupied by women and (3) economic status expressed as labour market participation. India however has scored a higher inequality score of 0.563 whereas the top 10 countries' score remained less than 0.05. This indicates the gross failure to deal with gender inequalities.**

In the flood that took place in Kerala in 2018 (referred to as the *Mahapralayam*), almost 10 districts out of 14 revenue districts were totally hit. As reported by the Government of Kerala, houses and property worth 19,500 crores were destroyed along with the sources of livelihood; more than 10,28,073 people were taken to relief camps, 483 people, 4 lakh birds, 18,532 small beasts and 3766 large animals were buried on known areas but this number could be much more if inaccessible forest areas could be reached. The floods washed away the freedom and aspirations of a good life of human beings. **Kerala is one of the few states in India with a consistent positive sex ratio (1084 females to every 1000 male) as reported in 2011 census. There is a high literacy rate in women, making Kerala distinct as compared to other states. But in a crisis situation it was seen that literacy alone cannot help in mitigating the impact of any disaster. In times like these, the capabilities, irrespective of class and education of people, count for resilience building of communities. Also, GDP and other economic measurements do not assure the well-being of women. When people are deprived of their basic capabilities, they will never be able to lead a good life.**

Major studies have shown that women are at a particular risk after a natural disaster. Deprivation, inequalities and discrimination, which are common features in pre-disaster society, only aggravate post-disaster. Some key findings from a study on women's mental health with reference to a case study of Uttarkhand (Parida 2016) highlight an unequal impact of disasters upon women:

1. Sanitation and hygiene was the biggest problem they were frustrated about.
2. The psyche of women was relatively more affected vis-à-vis men by the floods.
3. Majority of girl children drop out from studies after a disaster.

Also, the study on the impact of natural disasters on girls and women with reference to the earthquakes in Nepal (Norlha 2015) had brought out the following key aspects. They are:

4. Women, in particular single women, female-headed households, women with disabilities and older women, as well as girls are prone to immense dependence, economic and resource crisis after the disaster.
5. Women and girls go through a lot of stress, trauma, anxiety and frustration that is not attended to.
6. Trafficking and forced labour, combined with sexual exploitation, devastates the lives of many women during and post-disaster.
7. Some of their basic needs are simple to supply, yet governments and disaster management agencies are reluctant to take this forward.

However, UN Population Fund, formerly the UN Fund for Population Activities, (UNFPA) put together a dignity kit for Nepal which comprised of a gown, a sweater, a shawl, a petticoat, a toothbrush and toothpaste, a comb, underwear, a towel, a sari/dhoti, a reusable sanitary napkin, bathing soap and laundry detergent, a nail cutter and a flashlight with batteries.

Approaching Women in Flood-Affected Communities

The flood-affected communities were approached in a systematic way so that the voice of women and their conditions could be obtained. Personalized unstructured interviews were conducted in three stages:

1. Rescue Stage: from 22 rescue providers
2. Relief Stage: Contacts were established with 25 camp directors, 20 municipal councilors of Changanaserry, Alleppey and Kottayam, 18 panchayat members of Kottayam and Alleppey districts, 8 members associated with bar hotels, 55 members of NGOs like Rotary International, All Ladies League, Inner Wheel clubs and 15 individual social workers.
3. Rehabilitation/Rebuilding Stage: Around 545 flood victims who were moving out from camps back to their houses were contacted. These were in Kottayam, Kuttanad and Kumarakom. Discussions were held with district collectors of Alleppey and Kottayam. The study team held 2 radio interactive sessions with victims and social workers in Changanaserry and Alleppey and more first-hand information was collected.

Besides the above direct linkage with communities, much data was also collected from secondary sources, that is, government reports, research documents, books, periodicals and journals.

For the first stage, the entire flood-affected areas of Kerala was divided as per stratified sampling method into North Kerala, Central Kerala and South Kerala. Central Kerala was affected the most by floods; therefore, this was adopted as the area of study. The inputs received from various flood victims, elected representatives and district administrative authorities was drawn using Convenience Sampling (Fig. 1).

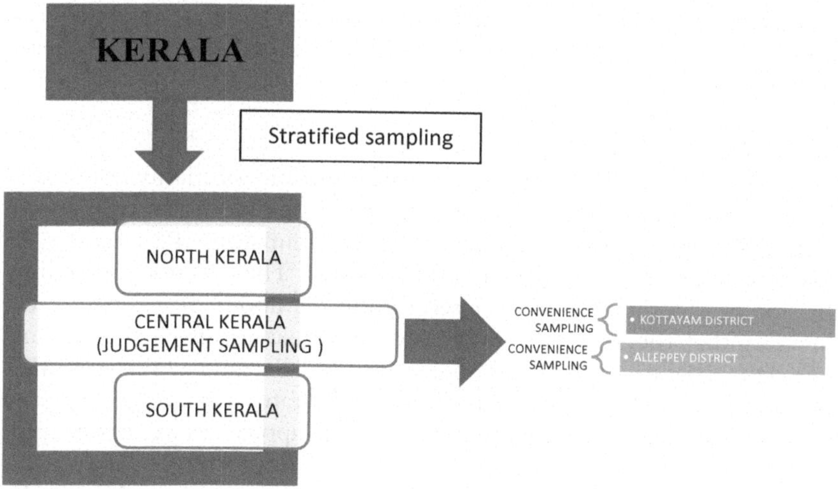

Fig. 2 Area-wise sampling process used in the study

Community Narratives

These are presented here as summing up of findings under a different name:

Case 1 (Women with Disability and Special Needs)
Jossymol, a 32-year-old unmarried, poor paralysed woman supported solely by her mother, lives with her parents in Kumarakom in Kottayam District, a place that is prone to floods every year. She is unable to stay in a relief camp as it is not disabled-friendly and she has special demands such as that of adult diapers and some other provisions. She is forced to live in a state of denial that the floods may not be as bad and she and her family prefer to stay stranded in their home during floods.

Case 2 (Women as Old Wives)
This case study is about the trauma of an elderly couple for whom the loss is irrtreivable in a short uncertain life ahead of them. Rescue workers found Marriama's husband in the interiors of Kuttanad, guarding whatever was left in his house and the only livestock that survived the floods was his pet goat. He requested them not to take him to the relief camp but carry his

goat to his wife who was in some camp unknown to him. Since they had lost everything, it was his keen desire that the goat, their only property, should be protected.

Case 3 (Old Age Homes)
Old age homes remain uncared for as the last priority of charity houses and the government. At the Jubilee Home, an Old Age Home for women run by the Snehagiri Missionaries Sisters in Chethimatom, Pala houses 55 elderly women and half of them are bedridden. Their compound wall by the side of Meenachil river collapsed during the floods. It was with great difficulty that they got two passers-by on a boat to lift all the patients upstairs to safer grounds. Their huge walls had fallen and all the crops that they had grown for sustenance had drowned. Water had also got into the home and all their valuable electrical household appliances were destroyed.

When the research team approached women during the floods, strange narratives emerged, which are still to go in the statistical data of damages and losses measurement strategies.

Dependence, fear of snakes and other risks:

They lived in a fear of death as water was rising and wild poisonous creatures such as snakes were entering their houses and their compounds. Evacuating people with disabilities was difficult and risky as water was rising. Living under mercy of others, camp directors, NGOs and government made people feel it was not worth living in their homes. In most of the houses, women who took the initiative to clean were exposed to life threats and having lost a lot of material possessions and other means of livelihood, some even went into depression. It was difficult to maintain hygiene relating to breast feeding. Nursing mothers wanted a little more privacy while nursing infants. Nutrition required for pregnant women was denied to them. Nursing mothers returned into their sullied and mud covered homes and it took a while to get all things clean. There was no support for being helped out.

Limited Access and Mobility to Collecting Food, Government Provisions
Some were starving for 3–4 days as their provisions were over. Food such as bread and biscuits were air dropped from helicopters but could not be eaten as their packaging was damaged and dirty water entered through the wrappers. No food of their own choice was given to them. However, in some shelters timely meals were given to them and some women who

were not used to evening snacks got used to the comfort of eating 4 meals a day. But this was not a usual and normal situation. In many interior areas, some women felt that they were deprived of ration and other necessities provided by the village offices on account of their political affiliations.

Women who were single or elderly and had decided not to live in makeshift camps but returned to stay in their neighbours' houses had to travel a lot and spend a lot of money in transportation, mainly boat services, just to get basic provisions. The government authorities never reached them and they were treated as non-existent citizens.

Inaccessible Medical Support

Most women had no access to medicines. Many middle-aged women needed medicines for high blood pressure and diabetes which were not available to them. There was great distress about the availability of sanitary pads for young girls and women and they also ran short of them in their homes. Many nearing menopause had excessive bleeding due to stress but no medical support was available. Besides, even the disposal of sanitary pads caused serious problems. When relief workers came to support it was embarrassing for the women to ask for sanitary pads. Many elderly and bedridden women using catheter or having wheezing or other health issues had a number of anxiety problems. Of many other medical problems, fever and gastritis were commonly reported. Visually impaired and physically challenged girls found it difficult to adjust to the new surroundings. Even though several medical camps were conducted, they were mainly general in nature and no women-specific camps were conducted. Much of the assistive equipment was lost in floods; wheel chairs, water beds and other supporting aids for the bedridden and physically challenged women were completely damaged and had to be bought again. A number of women reported urinary infections.

Indispensable But Unsafe Shelters

Ordinarily women in Kerala use their beds for storing their valuables like jewellery and material possessions. As their mattresses were drenched in water they could not use them. There were very few toilets and since benches or desks were used as beds, life in shelters was not comfortable. Some mats were provided to a few people as contribution came from NGOs. Many were also forced to stay in camps as the local government authorities had mentioned that only those registered in camps would get compensation post floods. Temporary shelters were made available to a few, but asbestos sheets and poor ventilation were some of the features which made life in shelters highly problematic. The state government delayed the assessment of houses and losses due to floods. There was also fear of assault, safety and security as locks and others security measures were not good and poor lighting in most camps in schools and colleges added to the misery. Pregnant women did not get required nutrition. Women who had made a choice to come to the camp for their own safety did not have the freedom to go anywhere in new surroundings. On the contrary, men visited bars in the neighbourhood and bought alcohol for themselves. Many women also reported that men were found smelling of alcohol and would play cards in the camps. Most women in shelters lived in fear of alcohol-drinking unbridled men in shelters.

Post Floods

Once water level came down, one could witness many relief workers plying on roads and in areas of Upper Kuttanad, a lot of women came to the main roads and stopped them to demand food and water. There was also a hoarding tendency amongst the better ones and this often obstructed relief workers from providing them with support. Post-flood delivery of essential goods was found to be discriminatory against women and also led to much humiliation for women who had to stand in queues for basic food items and women related needs shared in public.

Freedom, Equality and Justice: An Adaptation from the Flute Story of Sen in Idea of Justice

Questions which wait for answers are many. Who deserves to be saved first? Who deserves to be given preference in treatment relief and recovery measures? Who deserves justice first? *Women, Men, Children, Disabled, or the Old*?

To substantiate this, we take an illustration from the book 'The Idea of Justice' of the three children and a flute. We replace the flute with what is most important to each of the people in our three cases and who should be given in priority (Table 1).

In all the three cases women were deprived of their basic freedom because of the floods. The fulfilment of equality and achievement of justice can be obtained only when each is given what they want, in the measure they want and at the right time.

Table 1 Case analysis

Case	*Urgent requirement*	*Reason*	*Core capabilities*	*Freedom and justice denied as they were in the camp/house*	*Who would give them preference*
Number: 1 Jossymol, the paralytic girl	House needs to be remade as disabled friendly	This is not just to survive the yearly floods but also for an easy day-to-day life	Shelter Mobility	To choose different styles and ways of life	An economic egalitarian would feel she deserves it first as she was so poor and others more privileged than her
Number: 2 Mariamma, grieving the loss of her cattle	Livestock and cattle	She was grieving the loss of her cattle and livestock that used to give her milk for sustenance and livelihood	Other species	To grieve and express oneself	A utilitarian would feel she deserves it first as no one else knows how to use it to the fullest the way she knows
Number:3 Jubilee old age home	16-litre washing machine and compound wall	The sisters lost all electrical household appliances. For them a washing machine was the most important thing they needed as they had to wash 90 pieces of clothes daily as the elderly people soil their clothes. They cannot wash in river as it is dangerous. Sand mining has made the water deep and the compound wall that had broken added to the danger of going down to the river to wash. The three sisters' physical health would be at stake if they resorted to manually washing clothes. And if clothes were not clean, the bedridden elderly women could get all sorts of diseases and infections	Bodily health Control over one's environment: Material	To live a long normal life To be free from premature mortality To handle lives that were not part of their own nor could they handle their own lives	A Libertarian would sympathize here. After all, all their appliances that they had bought were a product of their hard work that had got destroyed

Findings

- Trafficking, forced labour or any form of sexual exploitation did not take place during and post-disaster in Kerala. This was suggestive of the fact that aware, literate and skilled women can protect themselves better than the women who measure less on these three assets. The camps were all safe due to the power and confidence of independent women.
- Unlike the 2013 Uttarakhand floods' findings, no girl child had dropped-out from studies in Kerala after the 2018 disaster. In fact, schools were cleaned once all left camps; books and other stationery were provided to the children, both boys and girls equally. Those who studied under CBSE syllabus and had lost their notes were given notes from a data bank that was compiled by a group of experts and students not affected by floods. Balance payment of school fees was waivered.
- Relief workers were much ill-coordinated, leading to several problems in distributing relief items they procured from NGOs and others. Misery of women was governance related and despite literacy and confidence they could not cross the threshold for obtaining support for women-specific items. Unlike the Dignity kit put together by UNFPA for Nepal women victims, there was no such standardized kit or procedure to distribute the same among the flood-affected women of Kerala.

Suggestions

1. Involve community NGOs like Kudumbasree for women in cleaning drives.
2. Incorporate Reduce, Reuse, Recycle and Say no to Plastic awareness trainings into the system. Educate them on hygiene and sanitation and share this with the local community.
3. Swimming has to be made compulsory in schools. If swimming pools are not available in the panchayat, funds should be set aside for make-shift pools, else a tie-up should be had with other schools or recreational clubs possessing swimming pools. Women too, irrespective of their age, need to be taught swimming.

4. Government needs to take disaster management more seriously, right from averting a disaster (precautionary measures) to quick response during disasters. Educating the community on the basics of alert and other things relating to evacuating and handling disasters is beneficial. Trial runs must be held occasionally.
5. Anganwadis need to be funded and replenished with toys and books to attract children. Seeds and saplings need to be distributed to houses so that the organic farming within their own home compounds can be reintroduced. This will also be beneficial for healthy living.
6. Elected representatives should be free of political bias. They should be free of politics when attending to the needs of their people. There is a need to introduce training sessions on the concept of 'Team building and Interpersonal Effectiveness' for the elected representatives of the local self-governments.
7. Both men and women need to be educated on women's rights, especially those relating to violence. For the sake of the women, men need to be administered to deaddiction programmes.
8. Camps should be made comfortable for all types of people to live. The propaganda that camps are safe should also be given.
9. Skill assessment should be done. This should be mapped with jobs or projects available. And women should be assisted in taking up a profession or job based on their skill sets. For this, government support is needed in financing such schemes.
10. Government can have a tie-up with NGOs like Archana women's centre and train women on masonry and carpentry so that they can first rebuild and repair their own broken houses and save money. They can also use these skills to build other houses and earn money.
11. Gender budgeting was introduced in Kerala State Budget in 2009, and later mentioned again with token amounts put in 2016–2017 and 2017–2018. All schemes relating to the rehabilitation of women can be routed through Gender Budgeting. There should be proper accountability for the same with a detailed report on the utilization of such funds in the following economic review.

Conclusions

In all the stages of disasters, women played the role of nurturer, caretakers by cooking food in their houses if possible while others went in search of food when their provisions got over.

Women were caretakers of the elderly parents as daughters and as daughters-in-law. Trained nurses made use of their professional skills and looked after people in the camps and in houses where they were stranded.

Women had to undergo humiliation and fearfulness as there was no such thing as a 'dignity kit separately given to women and girls'. The women reported that it was embarrassing for them to ask for their right size undergarments and sanitary pads. This was a lot of mental strain.

Women were also the un-supplied last mile for the government. Most of the women have mobile phones of their own and the young and middle-aged use whatsapp. However, they do not understand messages like red alert or orange alert that they receive post floods. Fear of not getting compensation post floods drove many to stay in camps. Unequal and unjust distribution of resources through the district administration system was a key feature due to 'favouritism' related to politics. This was prevalent in both districts under study. However, this was less seen in municipalities where women were in equal strength to men in numbers.

There was low attendance of children in Anganwadis where water had got in and educational material including toys were all destroyed. Mothers had a tough time keeping these children at home. It was mainly NGOs and lay people who organized relief materials. Many actively helped pets and animals as a large number of dogs were tied to pillars or gates and abandoned in the premises of their houses. Lack of empathy to pets was enormous, yet in the midst of all this perpetration of cruelty, one could find many women who stayed back to protect their livestock and other animals out of empathy as well as economic resource for the family.

References

Jahan, S. (2016). *Human Resource Development Report* 2016. Retrieved August 20, 2018, from file:///C:/Users/user/Desktop/jnu/2016_human_development_report%20capabilities%20as%20per%20un.pdf: http://hdr.undp.org

Kerala, G. o. (2018, February 22). https://kerala.gov.in/about-kerala. Retrieved October 8, 2018, from https://kerala.gov.in/

Ministry of Home Affairs. (2004). *Disaster Management in India.* National Disaster Management. Governemnt of India. Retrieved October 3, 2018, from https://www.unisdr.org/2005/mdgs-drr/national-reports/India-report.pdf

Nayagan, G. E. (2018). *Disaster Law Emerging Thresholds* (A. Singh, Ed.) New York: Routledge Publishers.

NORLHA. (2015). *Impact of Natural Disasters on Girls and Women Literature Compilation of Key Facts and Recommendations for Relief Practitioners with Regard to the Earthquakes in Nepal.* NORLHA. Retrieved October 1, 2018, from http://norlha.org/wp-content/uploads/2015/04/Impact_of_natural_disaster_on_girls_and_women_Norlha_June_2015.pdf

UNDP (2016). Human Development Report, Washington DC: United Nations Development Programme, Available at http://hdr.undp.org/sites/default/files/2016_human_development_report.pdf

Nisha Jose is a researcher in social and community resilience building policies, a social enabler and a motivational speaker. Her research inclinations include 'gender budgeting' in districts of Kerala.

Sony Kunjappan is an assistant professor at the Centre for Studies in Social Management, Central University of Gujarat, Gandhinagar.

Climate Change and Coastal Disasters of Bangladesh

Nasim Banu

Introduction

Bangladesh has a geographical area of 144,000 sq. km with 163 million population at the forefront of adverse impacts of climate change. Bangladesh is less than five meters above the sea level, located at the heart of the Asian monsoon region. Most areas of the country are also low-lying and flat due to which saline water inflow increases many environmental risks and impact of intensifying global climate change. Climate change-related disasters are impacting human systems, agriculture, forestry, fisheries, human and non-human habitats around coastal zones. Impact is severe on coral reefs, atolls, mangroves, boreal and tropical forest, ecosystem and native grassland (GoB 2010). The degraded environment and monsoon render Bangladesh highly vulnerable to natural hazards like flood, drought, cyclone storm surges and saline intrusion. The central and western coastal regions of the country are particularly vulnerable to cyclone storm surges and saline water intrusion. Severe cyclones hit Bangladesh almost every

N. Banu (✉)
Department of Development Studies, Islamic University, Kushtia, Bangladesh

A. Singh et al. (eds.), *Development in Coastal Zones and Disaster Management*, Disaster Research and Management Series on the Global South, https://doi.org/10.1007/978-981-15-4294-7_14

three years and the country bears about 40% of the impact of total global storm surges (GoB 2011). Due to cyclones with tidal surge, salinity intrusion devastates crops and livelihood of rural coastal regions. 'Bangladeshis', as the people of Bangladesh are called, are likely to face increasing scarcity of food and health provisions from climate change impact and their disadvantaged geographical location, combined with an extremely high density of population.

Coasts and islands are highly exposed to a variety of climate hazards (Westmacott 2002: 67). Increasing use and activities within the coastal zone, rapidly growing population, deteriorating environmental quality, loss of critical habitats, diminishing levels of fish and shellfish, reduced biodiversity and increasing vulnerability to natural hazards pose multiple threats to the county's economy. (Leary et al. 2009: 13). Consequently, people who live in the natural hazard coastal zones have had their homes and livelihoods destroyed in the past, with no safety net that could help them to recover. Effective disaster risk management in such vulnerable coastal areas of the country relies on a strong legal policy, inter-institutional coordination mechanism and, of course, community participation (IISD 2003: 3–7).

The core objectives of this chapter are to identify the impact of climate change upon the coastal people and to evaluate the country's ability to manage disaster risks. In order to meet the research objectives, comprehensive desk-based research has been done on the basis of public policy to climate change adaptation.

Impact of Climate Change in the Coastal Zone

The coastal zone of Bangladesh comprises 2.85 million hectares in area, 200 km in length, including 148,000 square km of rivers (GoB 2011). Out of 64 districts of Bangladesh, 19 southern districts with proximity to the Bay of Bengal come under coastal zone. The coastal zone of Bangladesh is a combination of 250 coastal islands, a large single tract of mangrove forest of Sundarbans, and a long sandy beach in Cox's Bazar, which is the habitat of more than 28% of the total population of the country.

Cyclones alone have caused the death of 250,000 people globally, among which 60% are from Bangladesh (UNDP 2004). Around 17% of cyclonic storm surges in the Bay of Bengal have landfall in coastal Bangladesh, while a severe one occurs almost every three years and causes loss of life, vital assets and livelihoods. During 2006–11, Bangladesh experienced US$ 2570 million in aggregate losses from 15 cyclones (Eskander and Barbier 2016). Climate change is expected to increase and therefore Bangladesh should be prepared to face and mitigate its consequences. Climate change threats are many and impacts people in many ways;

- Sea level rise with 45 cm is projected to inundate 75% of the World Heritage site, Sundarbans, the single largest mangrove forest in the world on which millions of people depend to earn their bread and butter (World Bank 2010). It has been said that the sea surface temperature of Bay of Bengal is steadily rising and might be related to an increase of cyclones as much by 10%, but there is no direct evidence available so far (G0B 2011). Cyclone tidal surge actions cause death, massive infrastructure damages, loss of livestock and others resources. Women and children become more vulnerable and face inhuman situations, forcing them to abandon their villages and migrate to safer pastures. This ultimately hampers the continuation of developmental works and education of coastal children.

The world's largest mangrove forest, Sundarbans, is a dynamically fragile and complex ecosystem in delicate balance with land and water. It is a natural coastal resource bank and a good habitat for wildlife, offshore fisheries, onshore shrimp cultivation and also highly valuable forest resources. This region faces phenomenal risk of erosion and denudation. Thus, for mitigating climate change effect on coastal people, the Government of Bangladesh (GoB) has put forth policy guidelines and increased public investment to manage disaster risks incorporating the elements emergency preparedness and risk mitigation, which is a marked shift from the earlier response-based approach.

Public Policies and Institutions for Climate Change Adaptation

The current disaster management practices of the country are committed towards the reduction of losses and damages to life, economics and environment. Over the years, GoB, has been giving equal importance to structural and non-structural mitigation measures, has invested in disaster risk measures and adopted various policies to protect the lives, livelihoods, low-lying agricultural land in the coastal belt from inundation and infusion of saline water and to make coastal people more resilient to climate change. The GoB has invested over USD 10 billion (UNDP 2004) in disaster risk measures which include:

- Construction of 2000 cyclone centres to provide refuge to the people affected by storm surges caused by cyclone;
- Construction of 6000 km coastal embankments and polders designed to raise agricultural productivity in the coastal zone and prevent it from tidal flooding and saline water intrusion (GoB 2011)

Structural mitigation measures are provided to protect the low-lying coastal agricultural land from inundation and infusion of saline water and have been so far been found effective. However, during the devastating cyclones of Gorkey 1991, Sidar 2007 and Aila 2009, these embankments could not ultimately resist the high and severe surge but got damaged and washed away in many places. Further, the water which entered the main land while receding damaged the embankment further. It is observed that places where the embankments did not have sufficient mangrove afforestation got easily damaged.

Starting in 1996, the government has been giving much attention to coastal afforestation which protects environment against natural hazards, stabilizes soil and makes the newly accreted land suitable for agriculture. The coastal forest region extends inland for about 120 kms, among which plantations cover an area of about 0.15 million hectares, mostly with mangrove spices. During the 6th FYP (2011–16) period, plantation had been done in a coastal area of 28,066 hectares; during the 7th FYP (2016–2020) period another 30,000 hectares will be planted/replanted in coastal zones (GoB 2016).

Legislation and Policies

As a non-structural part of mitigation, the GoB has been emphasizing meaningful legislation and policy, training and public awareness (Uthpal Kumar et al. 2009). To facilitate adaptation to climate change, over the years, the GoB has prepared and adopted various plan/polices to disaster and climate-reliant development initiative, pro-poor adaptation and mitigation, ecosystem-based disaster risk reduction and so on (GoB 2012). Bangladesh Climate Change Strategy and Action Plan (BCCSAP), 2009, aims to respond to climate change-induced development risk. The National Plan for Disaster Management (NPDM), 2010, aims to address multi-hazard risks and vulnerabilities. The Standing Orders on Disaster (SOD), 2010, has objectives to make the concerned persons aware of their responsibilities with respect to disaster management and accomplishing them. The Environment Policy and Implementation Program (EPIP), 1992, emphasizes on the conservation of natural resources like land, water, forest, and arresting salinity intrusion.

Besides the above, many dependent and subsidiary areas are impacted by policy implementation on climate risks and disaster management. National Tourism Policy (NTP), 1992, emphasizes the conservation of wildlife, particularly in Sundarbans. National Seed Policy (NSP), 1993, sparks the development of robust seeds for avoiding crop loss from natural hazards due to climate change. National Forest Policy (NFoP), 1994, emphasizes tree plantation along the coastal zones and manages the sustainable development of the Sundarbans. The National Fisheries Policy (NFiP), 1998, provides a hint of zoning the shrimp areas along the coastal belt. The National Agriculture Policy (NAP), 1999, emphasizes the use of ground and surface water for irrigation and post-disaster rehabilitation with emergency programme.

The Coastal Zone Policy (CZP), 2005, aims to (i) provide institutional arrangement for observing changes in climate of Bangladesh; (ii) assure maintenance of coastal embankment to table sea level rise; (iii) promote tidal basin management and planned coastal plantation; (iv) focus on pro-poor employment and capacity building for the poor on disaster risk management; (v) ensure saline-free and clean water supply along the coastal zone (GoB).

An Integrated Coastal Zone Management (ICZM), 2005, aims to create conditions for the reduction of poverty, development of sustainable livelihoods and the integration of the coastal zone into national process. ICZM can address multiple impacts and help to help community capacity for adaptation. The Climate Change Trust Fund of GoB finances the activities under BCCSAP, 2009, to make the country more resilient and less vulnerable to natural disasters (GoB 2009b). Bangladesh Climate Change Resilience Fund has been funded by several donors. Under this fund, Comprehensive Disaster Management Program has been initiated with very limited emphasis on coastal afforestation as a strategy towards the mitigation of cyclone and related storm surge impacts.

Training and Public Awareness

GoB has been training and disseminating information and skills needed for disaster management almost regularly to people and administrators in vulnerable zones. This is done through workshops, seminars, booklets, calendars and posters containing information about natural disasters. A disaster management chapter has been included in the educational curriculum of class V–XII. A compulsory lesson of at least 2 hours on disaster management has been introduced in the training curriculum of all types of training institutions of the country.

Institutional Arrangements

Beside policy initiatives for mitigation measures, there are institutional arrangements to address climate change-related disasters in the coastal zones of Bangladesh. These are:

- Bangladesh Meteorological Department (BMD)—generates warning regarding disasters like cyclone and storm surge and passes it on to public media and preparedness units for dissemination. On the event of the formation of a cyclonic storm in the Bay of Bengal, the Storm Warning Centre of BMD issues special weather bulletins from time to time till the landfall.
- Cyclone Preparedness Programme (CPP)—is an organization of volunteers at the field who mobilize people at the community level on how to mitigate and manage the impact of cyclones. The volunteers of CPP disseminate the warning signals, alert people, evacuate them to safe places, rescue marooned people after the cyclones, provide the injured with first aid, report on the losses and damages after preliminary assessment within twelve hours of the cyclone and so on.
- Water Resources Planning Organization (WARPO)—operates under the Ministry of Water Resources and is the focal point for water resources policy planning as well as targeted development of coastal zone of Bangladesh.
- There are some inter-related institutions created to ensure effective planning and coordination of disaster risk reduction and emergency response management such as (i) National Disaster Management Council, headed by the Prime Minister, (ii) Inter Ministerial Disaster Management Committee headed by the Minister, Ministry of Food,

Disaster Management and Relief (MFDMR), (iii) National Disaster Management Advisory Council, (iv) Ministry of Food, Disaster Management and Relief (MFDMR), (v) Disaster Management Bureau (DMB), and (vi) Directorate of Relief and Rehabilitation (DRR). There are broad-based Disaster Management Committees (DMCs) in the field levels at district, upazila and municipal/union headed by Deputy Commissioner (DC), upazilla nirbahi officer (UNO) and Chairman of Municipality/Union Parishad at respective areas.

Most policy measures mentioned above have shown policy successes. Following the devastating cyclones, disaster management activities including coastal afforestation have been upgraded by the GoB with technical and financial assistance of the UNDP and other donor countries and international agencies. These initiatives have been effective in dampening storm surge velocity during cyclone Sidr in 2007 and cyclone Aila in 2009, saving millions of lives and significantly reducing devastation as compared to the cyclone Gorkey in 1991; Gorkey caused 1,38.882 deaths while Sidr and Aila caused 3406 and 170 deaths, respectively.

The Integrated Coastal Zone Management (ICZM) Plan (2001–2005) was initiated by the Water Resources Planning Organization (WARPO) under the ICZM Plan, while the Water Development Board undertook a project entitled Coastal Embankment Rehabilitation Project for increasing the height of existing embankments and afforestation under the Department of Forest. ICZM Plan includes mitigation measures like (i) newly accreted land in the form of mud shall be utilized for planting mangrove plants, (ii) the existing mangrove forests shall be managed through participatory arrangements.

While reviewing the activities in progress, it indicates a lack of coordination and integration between disaster management activities of Ministry of Food, Disaster Management and Relief (MFDMR) and Ministry of Water Resources (MWR) with respect to mitigation measures for cyclones and storm surges. This aspect needs to be addressed in achieving synergy in disaster mitigation in the coastal belt.

Recommendations

A few key issues for sustaining and maintaining policy initiatives and efforts have emerged as recommendations:

- Regular maintenance of coastal embankments should be carried out so that in case of disasters such as tidal surges, the embankments can resist the saline water intrusion and protect the crops and lives.
- The content of the warning message should be straightforward and simple for common people to understand and not loaded with scientific processes and complex explanations. Messages should be clear and accurate with detailed information about the quantum of disaster that is likely to strike a particular area to help evacuation from critical areas to minimize loss of lives.
- The warning messages should be disseminated combining modern and traditional modes through both public and private, national and local and social media for the community as well as specific users.
- BMD should give special attention to occupational categories like fishermen for cyclone warning due to more vulnerability.
- Public awareness programmes/activities should be carried out throughout the year, particularly with regard to the role of forests and trees in cyclone impact mitigation.
- Attempts should be made to develop and increase volunteers at the community level under Cyclone Preparedness Program (CPP) with appropriate communication tools.
- Coastal land is fertile due to which people migrate towards coasts and settle in fringe land or islands. This calls for a properly regulated settlement policy on coastal zones land use in the coastal areas to reduce vulnerability.
- Structural and non-structural mitigation measures should be made an integral part of the development planning of the country to ensure investment in disaster prevention in the coastal zone. New economic activities can be ensured only through a robust and integrated mitigation plan.
- Impact study of cyclone mitigation and mangrove and sea grasses afforestation should be conducted in the coastal zones on a periodic basis with joint teams of plantation experts, disaster managers and social scientists.
- Increased investment should be made in early warning systems in generating community dissemination and public understanding as well as timely response from disaster preparedness organizations and personnel in cooperation with the likely affected community.

Conclusion

The observations made above point to considerable progress that has been made in cyclone disaster mitigation in Bangladesh as well as further efforts required in reducing loss of life and property through cyclones and related storm surges. Further activities are required for the integration of disaster mitigation-related activities of different departments as well as strengthening of institutional mechanisms through decentralization and local-level disaster planning. The planning process at the central level needs to recognize the interface between disaster and development. There is also a need to frame a high-level awareness and advocacy programme to create a better level of perception of disaster management including the aspects of long-term mitigation. Effective disaster mitigation measures are vital to the lifeline of coastal communities. Developing countries like Bangladesh with limited infrastructure and resources to cope with the impact of cyclones should receive particular attention of the international community in capacity building.

Reducing climate change effect for the coastal people of Bangladesh requires a sound plan and people's participation in community-based disaster management and technical improvement which can provide an early warning for the successful evacuation of people from vulnerable areas, thus resulting in fewer casualties. Many cost-effective planning, location, design and maintenance measures can be implemented which may reduce the risk and vulnerability of roads and infrastructure as well (Keller and MAsce 2002). Bangladesh is moving forward with plan/polices and institutional arrangements to protect the coastal people and make them resilient Therefore, development plan/policy would be the better approach for climate change adaptation for the coastal zone of Bangladesh.

References

Eskander, S., & Barbier, E. (2016). Adaptation to natural disasters through the agricultural land rental market: evidence from Bangladesh, Centre for Climate Change Economics and Policy Working Paper No. 266, May, UK: Grantham Research Institute on Climate Change and the Environment.

Government of Bangladesh. (2011). *The Sixth Five Year Plan: 2011–2015*. Dhaka: Planning Commission.

Government of Bangladesh. (2012). *Perspective Plan of Bangladesh 2010–2021: Making Vision 2021 a Reality*. Dhaka: Planning Commission.

Government of Bangladesh. (2016). *The Seventh Five Year Plan: 2016–2020*. Dhaka: Planning Commission.

IISD. (2003). *Livelihoods and Climate Change: Combating Disaster Risk Reduction, Natural Resource Management and Climate Change Adaptation in a New Approach to the Reduction of Vulnerability and Poverty*. IUCN, SEI, IISD and Inter Cooperation. Retrieved on January 15, 2009, from http://data.iucn.org/dbtw-wpd/edocs/2003-034.pdf

Kumar, U., Baten, M. A., Masud, A. A., Osman, K. S., & Rahman, M. M. (2009). Cyclone Aila One On, Unnayan Onneshan (The Inventors), 15/2, Indira Road, Dhaka, Bangladesh.

Leary, N., et al. (2009). For Whom the Bell Tolls: Vulnerabilities in a Changing Climate. In N. Leary, C. Conde, J. Kulkarni, A. Nyong, & J. Pulhin (Eds.), *Climate Change and Vulnerability* (pp. 3–30). London: Earth Scan.

The World Bank. (2010). *Bangladesh Country Assistance Strategy: 2011–2014*. Dhaka: Bangladesh Country Management Unit, South Asia Region, The World Bank Office.

UNDP (2004). Human Development Report, New York: United Nations, available at http://hdr.undp.org/sites/default/files/reports/265/hdr_2004_complete.pdf, accessed on 4th November 2019.

Westmacott, S. (2002). Where Should the Focus Be in Tropical Integrated Coastal Management. *Coastal Management, 30*, 67–84.

Nasim Banu is Chairperson of the Department of Development Studies and Professor of Public Administration in Islamic University, Kushtia, Bangladesh. She is a keen researcher in public policy and disaster management. Her focus is to assess the suitability of governance policies, planning and nature of public financing of Bangladesh to address the impact of climate change. She can be reached at: nasimbanu411@yahoo.com

Role of Insurance in Building Resilience for Coastal Zones: Market Versus the State

Shubhalaxmi Sircar

Introduction

There has been an unprecedented rise in weather-related events across the globe. Most countries have become increasingly vulnerable to the climate-related changes that are affecting millions of people and their livelihood. This is also leading to a rise in associated economic losses. According to a recent report by UNISDR, climate-related and geophysical disasters killed 1.3 million people and left a further 4.4 billion injured, displaced, homeless or in need of emergency assistance during 1998–2017. During this period, various countries reported direct economic losses valued at 2908 billion USD, and climate-related disasters contributed 77% of the total losses. There was a 251% increase in reported losses from extreme weather-related events between 1978 and 1997 and 1998 and 2017. The World Bank has estimated that the real cost to the global economy is 520 billion USD per year and these events are pushing 26 million people into poverty every year. Another alarming fact about the impact of disasters is that the

S. Sircar (✉)
Management Development Institute, Gurgaon, India
e-mail: subhalakshmi@mdi.ac.in

A. Singh et al. (eds.), *Development in Coastal Zones and Disaster Management*, Disaster Research and Management Series on the Global South, https://doi.org/10.1007/978-981-15-4294-7_15

worst affected countries belong to lower-income groups, led by Haiti. Therefore, the burden of disaster losses and destruction is borne by poor people around the globe. The Sendai Framework aims at building resilience of people and their assets and calls for a more comprehensive disaster management approach based on the proactive role of the State. One way of strengthening resilience and reducing risk of disasters is to provide insurance schemes. This shifts the discourse on how to tackle disaster preparedness from government funding to other forms of disaster risk reduction financing.

Objective

This chapter explores the various options available for financing disaster risk reduction and focuses on insurance as the method of risk reduction. The insurance sector is dominated by private players. Their objective is to assess the feasibility of insurance as a measure of disaster risk reduction through a market-based mechanism. This chapter seeks to compare insurance schemes available in developed countries such as the USA with the existing insurance schemes in India.

Definition of Disaster Risk

Disaster risk is defined as the likelihood of loss of life, injury or destruction and damage from a disaster in a given period of time (Global Assessment Report [GAR] 2015). The definition is self-explanatory: disasters are caused by natural hazards depending on the vulnerability and exposure of the people and the physical and natural environment adjacent to hazard conditions. Therefore, disaster risk can be defined as follows:

Risk = hazard * exposure* vulnerability

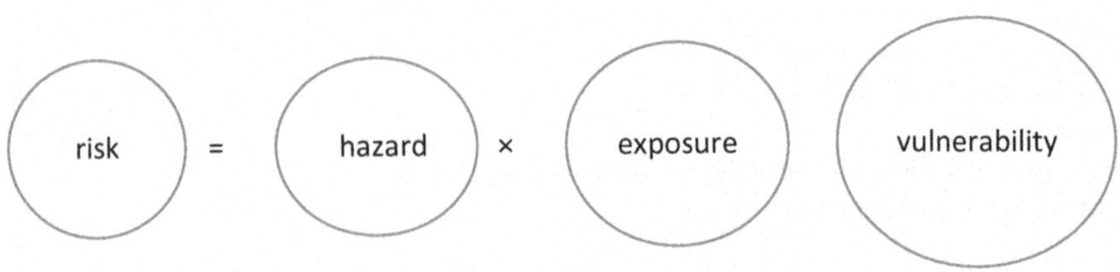

Source: https://www.preventionweb.net/risk/disaster-risk

Disaster risk is defined as '*the potential loss of life, injury, or destroyed or damaged assets which could occur to a system, society or a community in a specific period of time, determined probabilistically as a function of hazard, exposure, vulnerability and capacity*'(UNISDR, 2017) (report of the open-ended intergovernmental expert working group on indicators and terminology relating to disaster risk reduction. This report pertains to the development of a set of possible indicators for assessing global progress in implementing the Sendai Framework of Action 2015–30).

There are various ways in which disaster risk reduction efforts can be implemented and financed. First is national financing, where the majority of financing is done by the government both at the Central and State level. The other is international sources of financing. However, financing for disaster management is predominantly focused on disaster relief and rehabilitation. According to GAR 2015, disaster risk reduction (DRR) investments globally represent only 0.1% of the 6 trillion USD which will be needed for investing in infrastructure over the next 15 years. Moreover, it has been found that DRR investments have been beneficial for some countries for reducing disaster risk.

One can see from Table 1 that the benefits of DRR investments in countries such as China, Australia, Indonesia, the Philippines and Vietnam have outweighed the costs of investments made in these countries.

Table 1 Cost effectiveness of DRR investments in selected cases

Location	*Disaster type*	*Benefit/cost ratio*
Australia	All	3
Bangladesh	Cyclones, flood	1.18–3.04
China	All	35–40
Fiji	Flood	3.7–7.3
India Rohini River Basin	Flood	2–2.5
Indonesia	Flood	2.5
Maldives	Flood, tsunami, heavy rainfall, swell waves	0.28–3.65
Nepal	Drought	Expected average of 9
Pakistan-Lai River	Flood	1.3–25
Philippines	Volcano	>9
Vietnam	Cyclones	>4

Source: Financing disaster risk reduction for sustainable development in Asia and the Pacific, United Nations, ESCAP, 2015

There are various options available for financing disaster risk, and the options depend on the severity of natural disasters. For example, one of the options for high-severity disasters is international donors' aid. Government reserves and contingent credit can be used for high-frequency and low-severity disasters such as drought. Contingent credit facility is an ex-ante agreement which ensures available financing for disaster recovery and reconstruction. This is used in ASEAN countries like Lao PDR, the Philippines and Vietnam. The government can also set aside an allocated fund every year for frequent but small-scale disasters. However, government funds are not enough to mitigate financial requirements in the face of low-frequency but high-severity natural disasters such as earthquakes. What is needed is instruments which can transfer risk. That is, instruments are required such as insurance and CAT bonds available in capital markets. CAT bonds are essentially event-linked bonds which are used for payment for specific events. These bonds are issued by insurance companies to transfer a set of risks from issuer or sponsor to investors. Therefore, investors take on risks in the event of a catastrophe or natural disaster in return for attractive rates of investment. In case of an event occurring, investors lose the principal amount they have invested and the issuer (insurance or reinsurance companies) will receive that money to cover their losses.

Insurance products are one way of transferring risk. Several governments such as Japan's and New Zealand's provide insurance for disaster risk and support people with subsidies. In Japan, earthquake insurance schemes cover damage or losses from earthquakes, volcanic eruptions and tsunamis. These are supported by the government through a reinsurance scheme provided by the Japan Earthquake Reinsurance Co. Ltd.

The two most important objectives of any insurance scheme are:

1. to provide information to people living in hazard-prone areas such as coastal zones regarding the nature of risk they face;
2. to incentivize those people to invest in risk reduction measures prior to disasters.

The problem with the insurance market in developing countries is that the penetration rates are very low.

In developing countries, disaster insurance products are underdeveloped and insurance companies are often reluctant to provide cover for disaster risk. Traditional insurance schemes involve individual assessment of damage and loss. High operational costs and late payouts often make insurance products less popular. On the demand side, people do not have adequate financial capacity to pay for costly premiums. As a result, people living in disaster-prone areas prefer to depend on government reserves or contingency funds rather than buying costly insurance products. In many cases, the financial infrastructure is not sufficiently strong, leading to delayed payouts. Because of these limitations found in traditional insurance products, many countries in Asia have started using parametric insurance schemes. One widely used parametric insurance scheme is weather index insurance. Here, insurance is based on an objective parameter such as rainfall or temperature and payouts are triggered and calculated based on the agreed threshold level, for instance if rainfall is below a certain threshold in an area. This does not require individual assessment of losses and damages, which is not only time-consuming but also becomes quite costly. Hence, risk premiums are often very high. In a particular area, all policy holders receive the same payout, making parametric insurance quite cost-effective. Moreover, payouts can be made immediately as they do not rely on individual assessment of damage and losses, thus making them a very effective risk transfer mechanism. Insurance holders have direct access to when payouts are made and how they are calculated, thus making parametric insurance more transparent. Insurance schemes generally suffer from moral hazard problems. Insurance holders under parametric insurance do not have the incentive to indulge in risky behavior as the payouts are tied to an external objective which they cannot influence in any way, and they also know that this does not change the likelihood of receiving higher claims in any way.

However, parametric insurance also suffers from certain limitations. Parametric insurance payouts are estimated on the basis of an index which should be highly correlated with actual losses or damages. Otherwise, the actual losses can be overcompensated or are not adequately covered. It is often very difficult to find an index which accurately reflects the actual losses and damages, and hence this type of insurance is used for a limited number of natural hazards. Furthermore, it also requires sound technical knowledge to capture changes in environment so that the parameters can be accurately monitored. For this, even hourly and daily data are needed to calculate the payouts. In most developing countries, small and medium-sized insurance companies lack this kind of technical knowhow to design parametric insurance products.

Now we will look at the weather index insurance which has been used in the Indian agriculture sector. Indemnity-based insurance started in

India in 1999 through the National Agriculture Insurance Scheme (NAIS) and it was able to cover a meager 15% of farmers. The Weather Based Crop Insurance Scheme (WBCIS) has been the main publicly provided parametric insurance scheme, and was introduced in 2004. NAIS is a traditional form of insurance scheme which carries out an assessment for every insured farmer covered under the scheme. Under WBCIS, weather-related parameters such as temperature and humidity are used for payouts.

Insurance Program for Risk Mitigation in the USA

The National Flood Insurance Program (NFIP) was created in 1968 in the USA following the onset of Hurricane Betsy in 1965, and led to substantial federal relief for victims. At the time, there was inadequate coverage by private insurance players as the perception was that flood could not be adequately covered through market-based insurance schemes. The reason for this perception among private insurance providers was adverse selection because only particular areas were subject to risk: exceptionally high premiums would make the schemes unaffordable and in case of flood-related events the payouts would be so great in absolute terms that this would make a substantial dent in surplus or would lead to insolvency.

Florida imposed a moratorium on the cancellation and non-renewal of homeowners' insurance policies after Hurricane Andrew swept through the state in 1992. In November 1993, the state legislature enacted a bill that stated that the insurers could not cancel more than 10% of homeowners' policies in any county in Florida in one year and could not cancel more than 5% of their property owners' policies statewide for each of next three years. In 2007, Florida also sought control over rates and rolled them back based on new legislation that expanded the reinsurance coverage provided by the Florida Hurricane Catastrophe Fund (FHCF). Insurers were required to offer policies at rates below private market reinsurance rates to reflect the expansion of the coverage. The Citizens Property Insurance Corporation, a state-funded company in Florida, was formed in 2002 and controls a significant market share of the residential property market. Consumers are allowed to purchase a policy from citizens if a comparable policy would cost 15% more in the private market. This system encourages residents to move and build their homes in more risk-prone areas.

Coastal zones in Asia are extremely vulnerable to natural disasters such as tsunamis, hurricanes and flooding, and these events wreak havoc on lives and property due to burgeoning populations settling in coastal zones and river flood plains. Coastal areas of India have seen major hurricanes and resulting floods over recent years. Most of the population in developing countries such as India, Bangladesh, Indonesia and Thailand cannot afford the policies offered by private insurance companies as the premiums are too high. The burden of loss and destruction is borne by individuals, homeowners, farmers and small businesses. They have to depend entirely on government funds which are largely meant for reconstruction and rehabilitation and available only after the disaster has struck. The option of taking risk reduction measures to protect themselves and their assets is non-existent. There is no government-backed insurance program like the NFIP in the USA for people in developing countries residing in disaster-prone areas which are vulnerable to repeated occurrence of floods. They rely on their meager savings and assets to sustain themselves and their families in case of disasters. Recently, the International Water Management Institute (IWMI), with support from the CGIAR research program on climate change Agriculture and Food security (CCFAS), has created a pilot project based on the index-based flood insurance scheme with 200 farm households in the district of Muzaffarpur in Bihar. The project is used to provide compensation to flood-affected families during localized and mid-season flooding events in a time-bound manner, and also to ensure timely access to finance for smallholder farmers. The scheme is based on advanced modeling techniques using satellite data. This insurance product is fully subsidized and 14 farmers who suffered total crop loss received full insured amount of 20,000 INR per hectare.

Conclusion

India falls in the low-income group of countries, with per capita income much lower than other Asian countries. More than 50% of its population depends directly or indirectly on agriculture. India is extremely vulnerable to increasing climate change effects that cause floods and other weather-related events. As a result, people are more prone to disaster risk. It is thus very important for the government to start a subsidized insurance program which would help in building resilience and go a long way in strengthening disaster preparedness.

Shubhalaxmi Sircar is a senior faculty of Economics at the Management Development Institute, Gurgaon.

PART V

Case Studies

Coastal Flooding by Dam Mismanagement: Investigative Post-Disaster Study on *Criminal Negligence or An Act of God*

N. R. Joseph

Coasts Are Fighting Their Last Battle of Survival

For the last many years many international organizations and national agencies have been undertaking research on the increasing vulnerability of people and ecosystems in coastal habitats. Coasts around the world are facing severe erosion; sea level rise; destruction of natural mangrove forests, seagrass beds and coral reefs; and loss of marine species. Approximately 34% of the world's coasts are at high risk of erosion and this exceeds 69% in Asia alone (UNEP).[1] This increases the intensity and frequency of hurricanes, typhoons, floods, landslides and loss of livelihood. The global coastline is already much eroded[2] and subsequently coastal agriculture and livelihood is fast declining. In India the National Centre for Coastal

N. R. Joseph (✉)
Civil Engineer, Freelance Cross-cultural Structural Consultant and Architect, Mumbai, India

A. Singh et al. (eds.), *Development in Coastal Zones and Disaster Management*, Disaster Research and Management Series on the Global South, https://doi.org/10.1007/978-981-15-4294-7_16

Research under the Ministry of Earth Sciences has found that between 1990 and 2016, 34% of coastline had already been eroded.[3] The Report presents a highly threatening future for Indian coastal states as four main coastal states have lost nearly 50% of their coastal land to the sea, that is, West Bengal (63%), Pondicherry (57%), Kerala (45%) and Tamil Nadu (41%) (p. 17). The most threatening scenario is in Kerala, which has the most unstable coastal terrain (p. 15), due to which those who live on the coast have become highly vulnerable. The total loss of land at the coastal rim between 1990 and 2016 is 234.25 sq. kms (p. 17). This scenario leaves little scope for administrative negligence, apathy, corruption or misinformation. Repeated flooding along the Kerala coast indicates that decision makers need to distinguish between natural and man-made disasters with increased logic and clarity of analysis. The administrative luxury of apathy, negligence, misinformation and veiling the reality of climate change is detrimental for coastal people. The devastating disaster which this coastal state encountered in 2018 suggests the need for stronger accountability norms and regulations for authorities around coastal zones. Judicial intervention should be strengthened and new norms established for administrations to respond to peoples' voices with responsibility and respect.

Excerpts from Report Submitted to the High Court of Kerala by Joseph NR on the Floods of Kerala 2018 Being a Man-Made Disaster

Accountability cannot hide behind the 'Act of God' argument:

The courtroom battle to establish criminal negligence of the state government was challenging as a substantial data point and its appropriate proofreading cum comparative analysis ought to be undertaken with razor sharpness. The government had powerful data backed by hegemonic jargon, data, graphs and government memos in its Counter Affidavits. All the responding government departments reached the same conclusion against many PILs:

- The record 'Storm' (very heavy rainfall) of August 15, 2018 caused the flood;
- Contribution to flood from dam releases was minuscule.

These two arguments, if accepted, would quash the argument laid forth in the PILs that 'Criminal negligence caused the floods'. However, these arguments were not able to stand the test of scientific analysis supported by technology, field surveys and data analysis as provided by the country's top environmental monitoring institutions.

Main flaw in the government's argument

1. The government did not accept that the flood was already occurring by August 10, much before the August 15 storm rainfall. This was proven by the Chief Minister's public records, Government records collected and witnessed at television channels such as Doordarshan and Rajya Sabha TV, including many media reports. The sequence of events was produced before the court with sufficient evidence as given below:
   - July 20: Dams were at FRL/near FRL (Full Reservoir Level) in July as per affidavits that also admit *the levels were kept* (in breach of Dam Management Guidelines[4]) until August 8, 2018.[5]
   - August 8, 2018: Moderate rainfall raises water levels in reservoirs. Fearing overflow, shutters were opened by August 9, shown live on TV (as available at https://www.youtube.com/watch?v=kww6ddpKGls).
   - Dam spillage floods villages and towns much before August 15 and after August 8. This was demonstrated and displayed in several hundred reports and videos of this flooding before August 15, 2018. Government Affidavits are silent about this information even though it is all available online.
   - August 10: Kerala CM admits to floods and open dams in aid appeal (see Annexure 1) as follows:
     Our State is in the midst of an 'unprecedented flood havoc'. For the first time in history, 27 dams had to be opened. Never before had the State witnessed a calamity of this scale
   - August 12: Union Home Minister Mr Rajnath Singh visits Kerala, announces Rs.100 Cr Flood Relief Aid. Defense personnel join rescue efforts.
   - Rains are very low on August 11, 12, 13 and 14, 2018, incapable of floods (see the Indian Meteorological Department [IMD] rainfall chart) Yet houses, towns and villages in several districts were under five to ten ft. of water and over 110,000 people were

reported in Flood Camps. Where did the flood water come from? This was dam water flooding peoples' homes.

- August 14: Rainfall is low again but by midnight it suddenly intensifies into a storm. Where did all the water come from?
- August 15: Storm flooding is at full strength and gradually worsens.

The Finding

All Counter Affidavits filed by the government lost complete credibility by misleading the High Court, concealing information and producing the absurd stance that a storm can cause floods before it arrives.

Affidavits also confirm breach of cardinal Reservoir Guidelines, upheld up the Supreme Court in 2016, of dams NOT *maintaining low water levels during monsoons.*

Based on the above evidence, all of the following arguments produced in the Counter Affidavits by the Kerala State Government (KSG) to establish the flood disaster as an 'Act of God' were struck down null and void. The arguments which were struck down are as follows:

- KSG established the storm of August 15–17, 2018 as the sole cause of the August 2018 Floods (this implies the flooding began on August 15). KSG admits that water was released by dams but it was minuscule in comparison to the water brought in by the storm.
- KSG elaborated on the Record Storm of 1924 and its floods to equate this with the August 15 Storm. This was an erroneous comparison, as in 1924 there was only one dam and in 2018 there were 60-plus dams in Kerala, all of which blatantly bypassed dam safety guidelines and flood control measures in the management of dams.
- KSG claims that dam management was not at fault since the flood was caused by the record storm which was not predicted by the IMD. However, IMD's three hourly forecasts tell a different story.

Therefore the August Flood disaster was NOT an 'Act of God' BUT 'Man Made', demanding accountability and culpability on criminal negligence of KSG officials.

Dam Management and Safety of People Downstream

Every Affidavit admits that dam authorities in Kerala did not maintain low levels of water. This was in total disregard and breach of direct and clear instructions to keep water levels low, bearing in mind the monsoon in not just one but three guidelines by the following Central Government Authorities:

1. **Bureau of Indian Standards—BIS Guidelines—Indian Standard Operation of Reservoirs. 'Operation of Reservoirs' by Bureau of Indian Standards 3rd Para of Foreword (see Fig. 1)**

> *Control of flood is better achieved if the reservoir level is kept low in the early stages of the monsoon season. However, at a later stage, if the anticipated inflows do not result, the reservoir may not get filled up adequately for meeting the various water demands.*
>
> *On the other hand, if the reservoir is filled up to FRL in the early stages of monsoon, to avoid the risk of reservoir remaining unfilled at later stage, there may be problem of accommodating high floods occurring at later stage.*

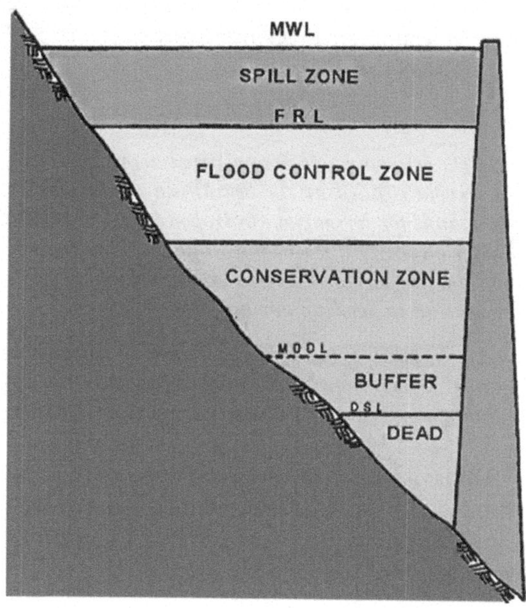

Fig. 1 Illustration in BIS guideline shows Flood Control Zone below FRL and not above it as mentioned in KSEBL affidavit. (Source: BIS Guidelines (2000) Indian Standard,Guidelines for Planning and Design of River Embankments, ICS 93.160, New Delhi: Bureau of Indian Standards)

> *Control of flood is better achieved if the reservoir level is kept low in the early stages of the monsoon season*
>
> *On the other hand, if the reservoir is filled up to FRL in the early stage of monsoon to avoid the risk of reservoir remaining unfilled at later stage, there may be problem of accommodating floods occurring at later stage*

2. **Central Water Commission—CWC Water Resources Ministry/ Dam Safety Procedures (see Fig. 2)**

Central Water Commission in 7.3.3 of the Operation & Maintenance of Reservoirs: '*In case of gated reservoirs, while it is desirable to fill the reservoir early, it should not be brought to near F.R.L* if late monsoon inflows are adequate to fill the reservoir*'.

7.3.1 *The aim of reservoir operation is to reduce the risk of man-made floods to the area downstream through carefully prepared reservoir regulation schedules*

3. **National Disaster Management Authority—NDMA Guidelines—Management of Floods. *Chapter 5 (5.4) on Regulation of Reservoirs* in the 2008 National Disaster Management Guidelines on the Management of Floods (2008) states:**

> *..if the reservoirs are not operated according to the operation manual and the reservoir is filled at the beginning of the monsoon season for meeting water demands for irrigation, hydropower, drinking and industrial water supply, a large quantity of water may suddenly have to be released to prevent the crossing of Maximum Water Level (MWL) and for ensuring the safety of the dam which may lead to flooding downstream*
>
> *The inability … to anticipate intense rainfall in the catchment, and consequent large inflows into the reservoir, may also necessitate the release of a large quantity of water resulting in large scale flooding*

Affidavits confirm that reservoirs were kept at and near FRL in August, a monsoon month, disregarding all guidelines. That is why even the moderate rain on August 8 and 9 2018 could raise water levels toward Over Flow, leading authorities to frantically spill heavily to avoid Over Flow

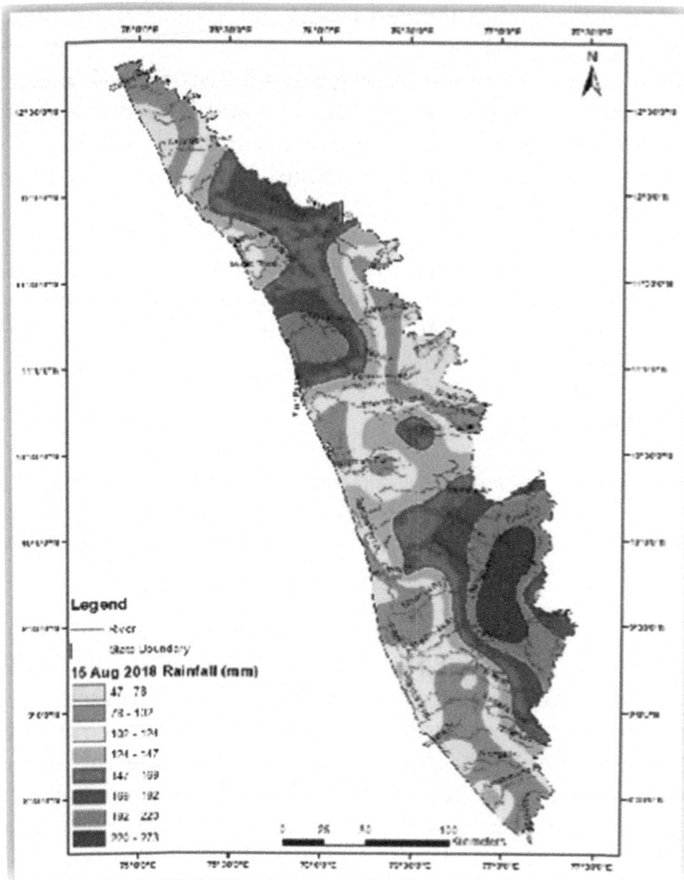

Fig. 2 The image shows the eye of storm of August 15–17 over Idukki (dark kidney-shaped area) where the storm was the strongest. (Source: Central Water Commission (2018) Study Report Kerala Floods of August 2018. Central Water Commission/Hydrological Studies Organization Hydrology (S) Directorate, New Delhi: Government of India)

And yet, Idukki was not flooded while many towns far from Idukki where the storm was not as heavy were flooded:

Chalakudy—under 10ft. water
Kuttanand—under 8 ft. water
Aluva—under 10 ft. water
Chengannur—under 8 ft. water

from 27 dams. This action flooded towns and villages along these rivers from August 9 to 19, 2018.

The Kerala Government also **disregarded a historic judgment of the Hon'ble Supreme Court.** Judges A.K. Sikri and R.K. Agrawal in Vohra Sadikbhai Rajakbhai & Ors Vs State of Gujarat 2016 delivered a verdict against the State of Gujarat which admonished the argument of the 'Act of God' as a creator of floods:

> *As per Respondents, the water had to be released from the dam as it reached alarming level because of heavy rains and non-release would have breached the dam. The action was, thus, taken in public interest and it was occasioned because of rains, which was an act of God.*
>
> *Verdict of the Court:*
>
> *Whether gross negligence in not maintaining particular level of water in the dam by the respondents; that has resulted into damage and destruction to the plantation of the appellants, causing loss of livelihood, could be said to be an "Act of God"?*

As per the PIL, the 2018 Kerala Floods were born on July 25, 2018 at an emergency meeting of the Kerala State Electricity Board (KSEB) officers when they decided to breach guidelines and do the exact opposite of what the Supreme Court verdict and the three pre-existing guidelines specified several times very clearly. Instead of lowering the water level, they decided to let the water rise to FRL (Full Reservoir Level) during August, a monsoon month, in all the dams even though they were already 85–98% full with July rain. This is what gave birth to the disaster, exactly as predicted in all the guidelines.

Given the unusually heavy monsoon in June/July, August rains *could not have* been small, *regardless of weather forecast*. Given below is the KSEB Order as action in blatant disregard of dam safety guidelines which gave birth to the August Flood. Dam authorities in an emergency meeting on July 25 decided *not to lower* water levels and instead allowed water to rise further (see Box KSEB Orders).

Box KSEB Order of the July 25 Meeting with the Minister who was the Electricity, Dam Safety and also the Disaster Management Minister. It is clear that the first alert was in itself a breach of guidelines:

The Chairman and Managing Director,KSEBL convened a meeting on 25-07-2018 to discuss about the matter of increase in water level at various reservoirs and the precautionery measures to be adopted by KSEBL. It was decided in the meeting to issue various ALERTs when reservoir water reaches the following levels in the case of major reservoirs.

Table-II

Sl No	Reservoir	First Alert	Second Alert	Third Alert
		BLUE ALERT	ORANGE ALERT	RED ALERT
1	Idukki	2390 ft.	2395 ft.	2399 ft.
2	Idamalayar	165 m	167 m	168.5 m
3	Pamba	983.5 m	985 m	986 m

Also it was decided to take the following measures during the First, Second and Third stages of alerts.

Normally no attempt to fill dams is made during the Southwest Monsoons that end in September. Following this, the Northwest Monsoon arrives and brings adequate water. When a moderate rainfall on August 8 and 9 came to endanger dams, dam authorities simply unleashed the water on the unsuspecting towns and villages downstream.

Therefore the KSEB Order is beyond sheer negligence in its nature in view of the fact that over 500 people died and over 21,000 crores was lost in damages. It was further treated as a criminal act holding billions of cubic meters of water, full to the brim, high above mountains below which hundreds of villages, towns and large cities were located.

Two Pertinent Questions Emerge

First, how was this storm responsible for heavy floods in Chalakudy, Parur, and so on when it did not flood Idukki where it was centered?

Second, what brought waters down to flood these riverside towns and cities where it did not rain as much?

Open dams flooded the rivers on the banks of which these town were located. This phenomenon is clearly indicative of the role of Reservoirs and Dam Management in the flooding that directed the waters from where it rained the most to flood areas where it did not.

How did the IMD Rain Chart help the 'Criminal Negligence' argument? Each line indicates average daily rainfall in Kerala in mm.

The tallest lines on August 15–16 are the **record storm.** The dense rainy days of July seen in the chart are what filled dams to 85–95% of their capacity after good rains in June.

Please Note: During the August 10–14 dates of the Flood Havoc the CM refers to, the rains are far below the 'Heavy Rain' marker. These rains were simply not capable of an unprecedented Flood Havoc.

Note that the spell of **Low Rains** between mid-July and mid-August occurred on the days that the dams could have released water in preparation for rain during August, a monsoon month, and kept the water levels low as sought by the guidelines.

The Counter Affidavit of the government alleged that the IMD forecast was wrong (Fig. 3). This argument is baseless and lacks scientific logic

Fig. 3 An analysis of the IMD Rain Chart 2018. (Source: IMD (2018) An analysis of the IMD Rain Chart 2018. Available at https://www.imdtvm.gov.in accessed on 7 Feb 2020

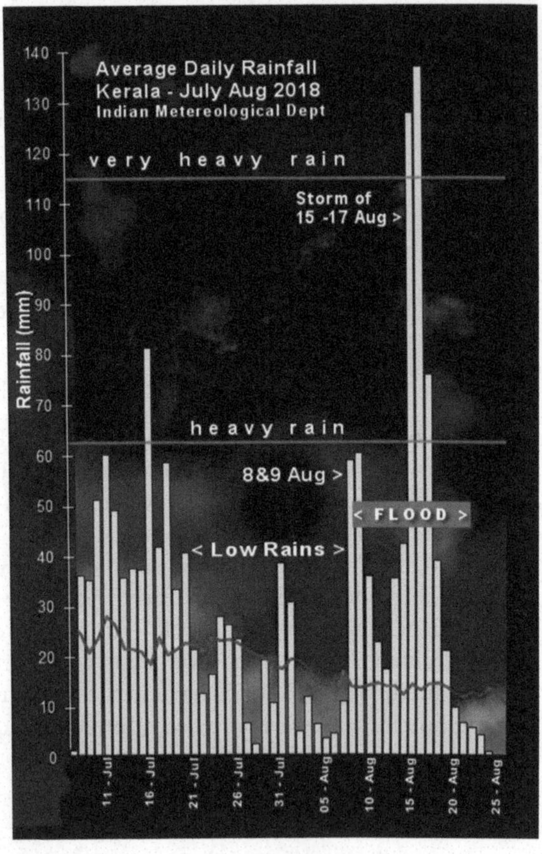

since weather forecasts are known to be based on stochastic probabilistic modeling (having a random probability distribution or pattern that may be analyzed statistically but may not be predicted precisely), inherently lack certainty and are expected to have high possibility of change. Every forecast also has a 'Forecast Probability' given alongside it and in the case of the forecast for August it is 47%, which means the forecast can go either way—the rain could get stronger or weaker. Forecasts always use terms such as 'Likely to be' and 'expected', which clearly indicate the nature of the prediction. All engineering systems that use these forecasts know this, from airplanes to shipping, sea ports and airports, and plan accordingly, aware that these are only probabilities given the large number of parameters that can alter forecasts at a moment's notice. Also, since the affidavit complains that forecasts for June/July were inaccurate, why did they expect the August forecast to be accurate? The 2018 monsoon had proven itself to be abnormal by June and again in July.

Ironically, if water levels were lowered even to accommodate the rain that the IMD had predicted in place of the storm, the flood would have been less severe as forecasts are given a good two weeks in advance.

'Sheer Negligence' or 'Criminal Negligence'?

By August 10, 27 dams in several districts were spilling 4.5 lakh liters/second into respective rivers. Large volumes of water released from 2700 ft. above sea level are extremely powerful and acquire high velocity. The water in the resulting flooding caused by these releases tears down embankments of rivers, roads, houses, bridges and so on, unlike rain floods into villages and towns on the banks of these rivers. These dams continue to spill, with many unable to shut down the flow immediately. Though the rains were moderate until August 14, it had built by August 15 into a record storm that wreaked even more havoc as flood levels in rivers rose 30–40 ft. to flow over highway bridges and submerge towns and villages on the banks of these rivers.

Could the August Floods Have Been Prevented?

Floods could have been avoided simply by following the guidelines and the Hon'ble SC verdict to lower water levels sufficiently during that July 25 meeting of the dam authorities. Computations show that if water level was maintained just 2.8 m below FRL, this would *have prevented flood* even during the storm of August 15–17.

The government's Counter Affidavits maintained that the water released from the dams was insufficient to have caused the floods. Now that it has been explained above that the flood had formed before the storm, this chart works against the affidavit as it shows clearly that discharge from dams was 50% equal to the amount of flood a once-in-a-100-year record storm brings.

As per chart Peak discharge = 8800 Cubic Meter/Sec
Discharge before peak on 14th = 2200
increase in discharge = 6600
Discharge Idukki/Idalamalar = 2750
+ from 14 others Dams = 440 (of 456 MCM capacity)
Discharge from Dams to flood = 3290% from Dams to Flood = 3290/6600 = **50%**

How is *50% contribution to flood* **minuscule**?

The very same graphs used by the CWC show how the dam releases were a significant 50% of the flood even during the record storm.

Similarly, the discharge from Idukki Dam also suggests and establishes that just by maintaining low water level in the dam, rain could have been contained in it and floods could have been avoided.

Water level in reservoir (**Green line**) rises to **FRL** due to steady inflow of rain water (**Blue**) at 500 cumecs. To bring water level below FRL they discharge water from dam (**Red Line**) which starts and remains at steady 750 cumecs discharge (horizontal portion of Red line). This action brings down the water level (Green dips) (Figs. 4 and 5).

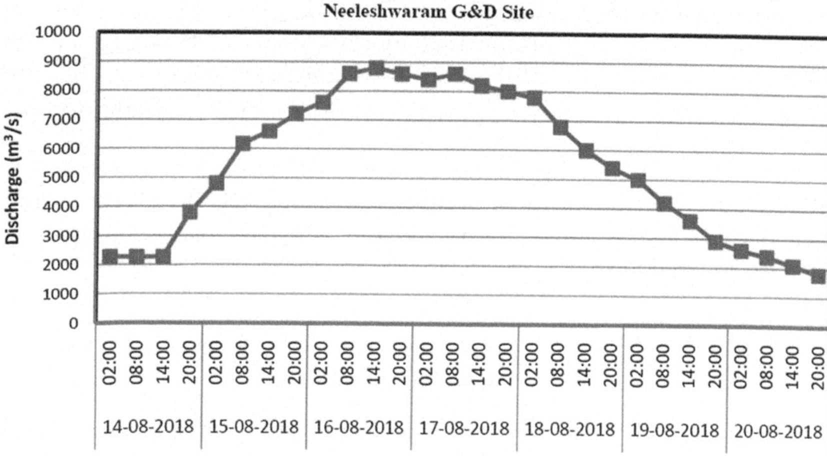

Fig. 4 Discharge data on Periyar River at Neeleshwaram G&D site. (Source: Study Report Kerala Floods of August 2018. Central Water Commission/ Hydrological Studies Organization Hydrology (S) Directorate, Government of India)

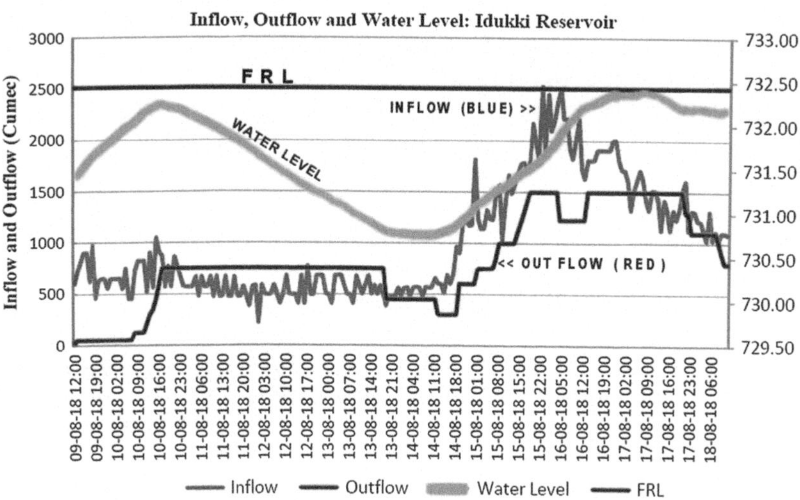

Fig. 5 CWC graph depicts struggle to keep water level that was just 1 m below FRL from overflow during storm in Idukki Dam. (Source: Study Report Kerala Floods of August 2018. Central Water Commission/Hydrological Studies Organization Hydrology (S) Directorate, Government of India)

Storm arrives on August 15. Water inflow (Blue) climbs up and so does water level (Green). Dam discharge is increased (Red goes up) to 1500 cumecs output from 500. This holds water level (Green) at FRL and eventually lowers it a little by August 18 as rain water subsides (Blue comes down).

Note: Had water level been maintained lower, at say 3 m (10 ft) below FRL instead of just 1 m in the 160 m-tall dams, **NO FLOOD would have occurred**. The entire drama above would have had enough space above to rise and fall.

EXHIBIT 15: HOW JUST 2.8 M COULD HAVE SAVED KERALA FROM FLOODS!

If Water Level Was 2.8 m Below FRL in August 2018 It Could Have Saved Kerala From Floods!

Consider Idukki Dam during the storm. The volume of storm water flow into the reservoir is the gray shaded area in the CWC graph, at 344 mill. cubic meters across four days. By spilling at 500 cumecs instead of 1500 that flooded Periyar, **44 MCM** could be discharged safely/day, storing only the rest (see Figs. 6 and 7).

Day 1 (86 MCM inflow less 44 spilled) = 42 MCM to store
Day 2 (129 MCM less 44) = 85 MCM to store
Day 3 (86 MCM less 44 spilled) = 42 MCM
Day 4 (43 MCM spilled in full) = 0 to store

Total to store = 169 MCM
Reservoir is 60 sq. km in area, **1 m** depth holds 60 MCM.
169 MCM would occupy 169/60 = **2.8 meters**
The story is similar with all other dams

Annexure 1(a) Appeal by CM of Kerala dated August 10, 2018 shows that flood existed before the storm and dispels affidavit's claims that storm created flood.

Annexure 1(b) Tweets from Chief Minister's Office Kerala on August 10—12 mentioning 'flood havoc' that clearly existed before the storm of August 15.

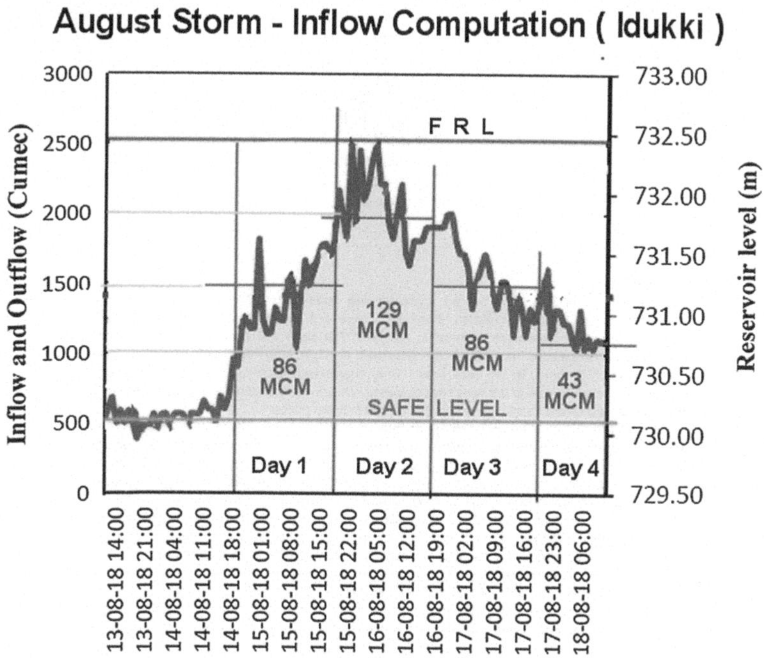

Fig. 6 Study Report Kerala Floods of August 2018. Central Water Commission/ Hydrological Studies Organization Hydrology (S) Directorate, Government of India)

Fig. 7 Chief Minister's Office Kerala appeal, Report: https://twitter.com/cmokerala/status/1028576388352110592?lang=en

GOVERNMENT OF KERALA
Pinarayi Vijayan
CHIEF MINISTER

10.08.2018

Dear Friend,

Our State is in the midst of an unprecedented flood havoc. The calamity has caused immeasurable misery and devastation. Many lives were lost. Hundreds of homes were totally destroyed and many more were damaged in the rains. The water washed away many shops and commercial establishments. The roads in the State were damaged on a large scale, with some roads being completely devoured by the flood waters. Even bridges had collapsed. Crop losses occurred in most parts. The misery due to the rains has not ceased. We are not able to make a complete estimate of the losses as the water is still above the normal level in many places.

For the first time in history, 27 dams in the State had to be opened. Never before had the State witnessed a calamity of this scale. In spite of that, we responded to the crisis quickly. We were able to rescue the affected and relocate them to camps. The entire state machinery worked in unison to provide drinking water, food and clothes to people in these camps. The services of Central Forces and Disaster Response Forces were made available without any delay. Army, Navy, Air Force, NDRF and Coast Guard are presently engaged in an exemplary rescue effort. The Police, Fire Force and other officials of the State are also doing a model service. Elected representatives and volunteers actively took part in the relief efforts. The Ministers coordinated the relief efforts at the district level.

Now we have an important duty before us, the task of bringing life back to normalcy. The people of the State must come together for this effort. The leader of the opposition was also a part of the official team of Government that visited the affected areas. The situation demands from us a unity that would make us proud forever. The Union government has promised all possible assistance to the State.

We become civilized only when we show compassion towards those who are suffering. We came together during the time of Ockhi disaster. In a similar manner, we must realize that this is a calamity of the entire State. I urge all to consider this as a request of Kerala and contribute generously to relief efforts.

Yours sincerely,

Pinarayi Vijayan

Address to mail Cheque/Draft:	*Account details for transferring online:*
The Principal Secretary (Finance), Treasurer, Chief Minister's Distress Relief Fund, Secretariat, Thiruvananthapuram – 1.	Account Number: 67319948232 Bank: State Bank of India Branch: City Branch, Thiruvananthapuram IFS Code: SBIN0070028 PAN: AAAGD0584M Name of Donee: CMDRF

Contributions to the Chief Minister's Distress Relief Fund (CMDRF) are 100% tax exempt under section 80G of the Income Tax act.

141, 3rd Floor, North Block
Government Secretariat
Thiruvananthapuram Pin 695001

T 0471 2333241, 2333812, 2333610 F 0471 2333489
E chiefminister@kerala.gov.in
facebook.com/cmokerala twitter.com/cmokerala
www.keralacm.gov.in

Residence
Cliff House
T 0471 2318406, 2314853

https://twitter.com/cmokerala/status/10279045083014348 80?lang=en

CMO Kerala @CMOKerala 10 Aug 2018 (Fig. 8)

Kerala is in the midst of an **unprecedented rain havoc.** Path to recovery can be long & arduous, but you can make a difference by joining the rebuilding efforts. CM Pinarayi Vijayan has urged all to contribute generously to the Chief Minister's Distress Relief Fund. #KeralaFloods

Fig. 8 Flooded Aluwa/Kochi area due to Idukki Dam release on August 10 when the rain was just 20 mm (low) (aerial view) The water level at Idukki Dam was 2398 at 8 am, against the FRL of 2403 feet. Idukki Dam gates in Kerala were opened after 26 years. (Source: DD News (2018) 25 August 2018, retrieved from YouTube site https://youtu.be/-YBbAfw1Ka4)

This image was uploaded on August 10, 2018, and questions dam management: https://youtu.be/-YBbAfw1Ka4. On August 9, 2018 flooded streets and houses at Kalpathy in Palakkad, after Malampuzha Dam released water having held water at FRL for days (https://www.youtube.com/watch?v=PKFYWrKaaWg).

Link: https://www.thehindu.com/news/national/kerala/red-alert-issued-for-eight-districts/article24657832.ece

The author gathered information from people as a primary source of knowledge. It was found that 'Dozens of flood victims interviewed by Reuters, who live in villages dotting the banks of Kerala's biggest river, the 244 km Periyar, say *they faced no floods despite torrential rain in late July and early August. All of them said waters rose overnight Aug.15.*That was when more intense rainfall forced KSEB to rapidly ramp-up releases of water from Idukki and Idamalayar reservoirs, which feed into the Periyar'. https://sandrp.in/tag/kerala-floods/

Conclusion

Most disasters are avoidable if an appropriate, sensitive and alert decision maker works efficiently with a conscience to go beyond the call of duty. The 2018 Kerala Floods were avoidable and were not the result of natural causes termed an 'Act of God'. Evidence produced before the Hon'ble High Court establishes that the disaster was caused by sheer 'criminal negligence'. However, lack of accountability norms in governance have maintained the conditions and officials which caused these devastating floods. This chapter indicates a lack of will toward strong disaster management in the country.

Acknowledgments Special thanks to https://sandrp.in/ and Professor Amita Singh, Chairperson of the Special Centre for Disaster Research and the Series Editor for this volume, for support and encouragement in the writing of this chapter.

Notes

1. See UNEP (United Nations Environment Programme) at https://www.unenvironment.org/explore-topics/oceans-seas/what-we-do/working-regional-seas/coastal-zone-management.
2. At present, coastal erosion is very common world-wide, and approximately 70% of the world's sandy coasts are marked by erosion. Coastal land is being swallowed by seawater and coastal villages and houses are forced to move inland, which squeezes the living space of humans, destroys beach biodiversity and ecological balance, and has direct or indirect harmful effects on human life and the natural environment. Thus, coastal erosion has been transformed from a natural environment change to a serious hazard (Yincan et al. [2017] 'Coastal Erosion', Chapter 7).
3. Kankara, et al. (2018) *and* Annual Report 2018–19, Ministry of Earth Sciences.
4. The Supreme Court in 5 holds the guideline to maintain low water levels well below FRL during monsoon as a cardinal caution that dams must follow. Every Affidavit admits that dam authorities in Kerala did not maintain low levels of water. This was in total disregard and breach of direct and clear instructions to keep water levels low, bearing in mind the monsoon in not just one but three guidelines by the following Central Govt Authorities, that is, (1) Bureau of Indian Standards—BIS Guidelines on Indian Standard Operation of Reservoirs; (2) Central Water Commission—CWC on Water Resources Ministry/Dam Safety Procedures; (3) National Disaster

Management Authority—NDMA including Chapter 5 and 6 on Regulation and Enforcement related to flood control and dam safety, National Disaster Management Authority (2008).

5. Flood relief announcement of (Rs.100 cr) was made on August 12, 2018, before the storm by the Ministry of Home Affairs and the State Chief Minister's official tweets on 'Flood Havoc' were online on August 10–12 before the storm; also see article at https://timesofindia.indiatimes.com/city/thiruvananthapuram/half-of-kerala-reels-under-floods-as-29-die-54000-are-homeless/articleshow/65362118.cms

References

Annual Report 2018–19. *Ministry of Earth Sciences, Earth System Science Organization.* New Delhi: Government of India.

Kankara, R. S., Murthy, R., & Rajeevan, M. (2018). *National assessment of Shoreline Changes along Indian Coasts-A Status Report for 26 years 1990–2016.* Chennai: NCCR Publication. Website: http//www.nccr.gov.in.

Vohra Sadikbhai Rajakbhai & Ors Vs State of Gujarat (2016). Supreme Court Civil Court Appeal No:1866 of 2016.

Yincan, Y., et al. (2017). *Marine Geo-Hazards in China* (pp. 269–296). Amsterdam/Oxford, UK: Elsevier. Online: http://www.sciencedirect.com/science/article/pii/B9780128127261000073.

Online Resources

Bureau of Indian Standards. (1994, November). *Indian Standard: Operation of Reservoirs.* https://ia800204.us.archive.org/12/items/gov.in.is.7323.1994/is.7323.1994.pdf

Central Water Commission (CWC) / Ministry of Water Resources / Dam Safety Procedures http://cwc.gov.in/main/downloads/Report%20on%20DS%20Procedures.pdf

Dam induced/Role of dams in Kerala's floods https://sandrp.in/2018/10/04/role-of-dams-in-keralas-2018-floods/#more-30696

Did Dams make India's Once in a Century Flood worse? / Reuters. https://widerimage.reuters.com/story/did-dams-make-indias-once-in-century-floods-worse

Idukki Dam releases water when Kerala is in Floods: Could this have been avoided? https://sandrp.in/tag/kerala-rainfall/

National Disaster Management Guidelines / Management of Floods / Jan 2008/https://ndma.gov.in/images/guidelines/flood.pdf

Study Report: Kerala Floods of August 2018. Central Water Commission (2018, September): http://cwc.gov.in/main/downloads/KeralaFloodReport/Rev-1.pdf

Video BBC News / https://www.youtube.com/watch?v=sJ5-HCYL2Hw

Video – Compilation of Twitter clips on Kerala Floods. https://www.youtube.com/watch?v=MXWGhs6lW5w
Video: Kerala Floods – The Complete Picture / Manorama Online. https://www.youtube.com/watch?v=-kHEI_76Bnc
Video / Kerala Flood: The Human Story in English II Documentary By Discovery 12 Nov 2018. https://www.youtube.com/watch?v=nXtx7SdM65M
Video TV18 / Monsoon Fury kills 100 s. https://www.youtube.com/watch?v=DbaGsIJAlwY
Why Kerala floods proved so deadly / 21 August. https://www.bbc.com/news/world-asia-india-45243868

N. R. Joseph Petitioner in First PIL before Kerala High Court that the 2018 floods of Kerala were caused by criminal negligence by dam authorities

(N. R. Joseph obtained a Civil Engineering (NIT Trichy, 1980) and a Masters in Industrial Design (IIT Mumbai 1982). Works as a freelance structural consultant, architect (Bangalore ISKCON temple is his best known work) and as a multimedia creator/consultant. Has executed works for Kochi Refineries, Cochin International Airport Limited, Dubai Shopping Festival, etc.)

Coastal Ballads and Conservation Ironic: Understanding Implementation Slippages of the CRZ Law

Amita Singh

The passage and progress of the Coastal Regulatory Zone law from 1991 until today has carried it through a number of environmental and developmental challenges. This law applies to the conservation of fragile ecology and ecosystems surrounding all water bodies such as rivers, creeks, lagoons, estuaries, coral reefs, mangroves, swamps and backwaters. The need for considering environmentalists of the stature of Madhav Gadgil and the Kasturirangan report for preparing a report on the Western Ghats and then succumbing to the populist resistance which followed against their recommendations suggests a need to raise one pertinent question before expanding development into fragile eco-zones: Can the fragile

This paper was published as a Special Article in Economic and Political Weekly, Vol. 51, Issue No. 7, 13th Feb. 2016 and is being reproduced here with an authorized permission from EPW.

A. Singh (✉)
Center for the Study of Law & Governance, Jawaharlal Nehru University, New Delhi, India

A. Singh et al. (eds.), *Development in Coastal Zones and Disaster Management*, Disaster Research and Management Series on the Global South, https://doi.org/10.1007/978-981-15-4294-7_17

ecology of riverbeds and coasts be preserved without substantive land-use restrictions over them? This chapter attempts to find answers to this question and to demonstrate that the gap which is created due to state failure in acting as a custodian of 'environmental resources' has placed the judiciary in a powerful position with immense freedom to interpret the CRZ regulations. This has weakened the spirit of law by reducing the scope and effectiveness of regular administrative agencies expected to implement the law and conservation requirements. This chapter highlights that any wavering on implementing CRZ law will push the fragile vicinity of water bodies into increased vulnerability to disasters, leading to massive socio-economic destruction and loss of lives.

What Is a Coastal Zone Land and the Land Called Riverbed?

A coastal zone is the land area from the low tide line in the sea waters to the area that the sea water reaches during high tide. This covers the land around rivers, creeks, lagoons, estuaries, coral reefs, mangroves, swamps and backwaters which is affected by tidal action. Coastal zones provide and sustain diverse ecosystems which produce goods and services that can never be substituted even by the best government provisions. The United Nations Ecosystem Assessment Report of 2005 provides a detailed study of the value of these goods and services to human beings. In 1993 a Writ Petition (No.664 of 1993, decided in 1996) from the Indian Council for Enviro-Legal Action raised a grievance against the Central Government for not enforcing the implementation of CRZ Regulations by state governments, leading to 'continued degradation of ecology in the coastal areas'. The Writ Petition also challenged the validity of the 1994 Notification which sought to amend and 'defeat the intent of the main 1991 Notification'. Following this, the Hon'ble Supreme Court directed the states to set up State Coastal Zone Management Authority (CZMA) which would formulate a Coastal Zone Management Plan (CZMP) to implement CRZ Regulations. This directive from the court demonstrated that this land area of coasts and riverbeds was different from the rest of the district land area and is governed by a different set of rules. It also indicated that the special and exclusive attention which this land area requires may not be within the capacity of district land monitoring and developmental agencies such as the District Town and Country Planner or the Municipal Corporation. The states were monitored for compliance and the Chief Secretaries of the non-compliant states were issued notices to explain and show cause. This strict action by the Supreme Court led the states to activate conservation efforts in the CRZs.

Riverbed area has been a politically contested zone. A legitimate understanding emerged[1] on the issues raised by Justice Gita Mittal in the Case of Commonwealth Games Village[2] in 2009, which stated, 'The river, riverbed, river basin, the flood plain of the river is not within the meaning of the expression land and consequently its change of user is not permissible'. The learned judge referred to earlier orders[3] of May 2005 and 2003 to emphasize that '…river Yamuna its bed, basin and flood plain is a water body, is not land and can be utilized only as a water body'. However in 2009, a three-judge Supreme Court bench of Justice B.S. Chauhan, K.G. Balakrishnan and P. Sathasivam reversed this normative and ecological understanding to a new institutional explanation quoting scientific literature and global opinions which do not treat riverbeds as a water body but as land which can be constructed upon. Similar fuzziness was created when builders started occupying biodiversity-rich spaces of backwaters for constructing holiday resorts. There is no mandatory ecological or sensitization training of judges and bureaucrats managing the environment as they advance in their institutional hierarchy, consequently restricting the scope for ensuring environmentally sound and sustainable development in the country.

Land Use in Fragile Areas

The protection of the fragile[4] ecology of riverbeds and coastal zones makes land-use restrictions indispensable. The National Centre for Sustainable Coastal Management (NCSCM), which comes under the Union Ministry of Environment, Forest and Climate Change, recently concluded a study on India's coastal zones to highlight that in the last 50 years more than 40% of the coastal area has eroded.[5] Repeated and frequent climate change-related disasters in the form of floods, hurricanes, cloudbursts and landslides have increased ecological and environmental vulnerability, which, when coupled with incessant, ambitious and unregulated urbanization, destroys sustainable ecosystems forever. Water bodies such as lakes, ponds and underground aquifers have become garbage dumps, construction sites or golf courses. However, this serious policy failure has repeatedly been ignored and rejection of the Gadgil Committee Report on the Western Ghats only adds to the list.

The need for land-use restrictions around coasts, rivers and other water bodies led to the Coastal Regulation Zone Notification 1991. In exercise of the powers conferred by Clause (d) of sub-rule (3) of Rule 5 of the Environment (Protection) Rules 1986, the Central Government declared the coastal stretches of seas, bays, estuaries, creeks and backwaters which were influenced by tidal action (in the landward side) up to 500 m from the High Tide Line (HTL)[6] and the land between the HTL and the Low Tide Line (LTL) as the Coastal Regulation Zone and imposed restrictions on its use for setting up industries and other operations. An incremental approach was used to restrict activities on the landward side of the HTL. The regulation of activities over land was ensured through a four-fold classification of CRZ: CRZ-1, CRZ-2, CRZ-3 and CRZ-4. The Notification further added that no construction (including temporary constructions and fencing or other such barriers) would be permitted within 200 m (in the landward side) of the HTL and also within the area between the LTL and HTL. This area is designated as a 'No Development Zone' (NDZ).[7] The declaration of the NDZ brought down the guillotine over hoteliers and industrialists who had become more ambitious due to the global flow of capital and professionals into India. Therefore amendments made to the CRZ Notification in 1994 exempted new construction and developmental activities in the NDZ. The Notification also reduced the distance from 100 m of the HTL to 50 m. If this was allowed to move further it would have caused massive devastation but due to an alert and timely petition in the Supreme Court by the Indian Council of Enviro-Legal Action vs. Union of India and others[8] this legal overreach was immediately struck down.

The 1991 Notification has, however, regulated land use for developmental activities up to 500 m of the HTL on the landward side. Annexure I to the Notification deals with coastal area classification and development regulations, but it also draws attention to the fragile ecology of these coasts, as in the case concerning Vembanad Kayal backwater[9] adjoining Vettila Thuruthu and connected lagoons and filtration ponds in which the Kerala High Court referred to the area as a Critical Vulnerable Coastal Area (CVCA). This generated increased attention of courts toward land utilization restrictions for the greater public good.

The biggest hurdle which courts encountered in imposing land-use restrictions came from an argument on the loss of livelihood. CRZ is an area which generates livelihood for local inhabitants by providing goods and services to the whole region. This area serves as a common property resource of villages which provides free ecosystem services benefiting the poorest of the poor, who survive in close proximity to nature. Some of the estimates emerging to calculate the cost of ecosystem services have confused those who thought it was a free asset. The study highlighted that:

> We have estimated the current economic value of 17 ecosystem services for 16 biomes, based on published studies and a few original calculations. For the entire biosphere, the value (most of which is outside the market) is estimated to be in the range of US$16–54 trillion (1012) per year, with an average of US$33 trillion per year. Because of the nature of the uncertainties, this must be considered a minimum estimate. Global gross national product total is around US$18 trillion per year. (Costanza et al. 1997: 253)

Thus the price of ecosystem services was calculated as almost the double the total world economic output in 1997 when Costanza and colleagues attempted to estimate this amount. More recently in 2012, John Dearing's (2012) study highlighted the deleterious effects of human-induced land-use and land cover changes leading to the transformation of landscapes and the greatest loss to biodiversity and environmental balance. The study strongly demands robust action against unchecked development.

The 12th Plan objective of investing in the prevention and mitigation of environmental disasters with a focus on resilience building of ecologically fragile and environmentally sensitive regions suggests a step toward recognizing the vulnerability of local populations. Communities inhabiting ecologically sensitive areas such as fisherfolk, Sabai grass workers, and forest or mountain tribes are becoming increasingly vulnerable due to intrusive development and irregular construction in their areas. The need for livelihood security is linked to conservation in para 1 of CRZ Notification 2011 where reference has been made to sub-section (1) and clause (v) of sub-section (2) of the Environment (Protection) Act 1986, through which the Central Government has been conferred the responsibility of ensuring sustainable livelihood security for fisher communities and other local communities living in the coastal areas. In doing so, it legitimately imposes many land-use restrictions such as the setting up or expansion of industries; operations, processing and manufacture, handling or storage, or disposal of hazardous substances as specified in Hazardous Substances (Handling, Management and Transboundary Movement) Rules 2009; or any construction activity. In many instances the Supreme Court has even set aside High Court permissions for construction which violated the CRZ conditions for land-use diversion[10] and appointed Expert Committees to prepare an Integrated Island Management Plan for an appropriate and sustainable development of the area.

Since environmental protection is recognized as a fundamental right, environmental degradation becomes a violation of the fundamental right to life under Article 21 of the constitution. On several occasions since its most explicit approval in the Dehradun Quarrying Case (AIR1988, SC 2187), even the judiciary has admitted that the right to a wholesome environment is indispensable for sustainable human well-being. This has subsequently featured in the judgments delivered in the infamous Oleum Gas Leak Case (AIR 1987, SC1086), Arvind Textiles v. State of Rajasthan (AIR 1994 Raj.195) and AP Pollution Control Board v. M.V. Nayudu (AIR 1999, SC 812). Thus environmental protection in an eco-sensitive region such as backwaters creates heightened responsibility for both the Central and state governments toward greater fairness and eco-centric conservation.

CRZs protect and sustain livelihood as much as the coastal ecosystems do. In going back to the history of CRZ regulations one would situate the National Fishworkers Forum (NFF) as the pioneers of this conservation milestone. Distressed with depleting fish catch due to industrial, oil and sewage pollution, combined with ambitious dredging and reclamation activities, the NFF marched through Kanyakumari in 1989 demanding pollution-free coasts. Their slogan was 'Protect Water and Protect Life' as they highlighted the link between all water bodies from the sea to the lakes and rivers. It was then that the government responded by bringing in the 1991 CRZ Notification. The recent defiance of CRZ by builders and local authorities has once again brought into focus the NFF under the well-coordinated platform of the Kerala Swatantra Malsyathozhily Federation. Their slogan throughout the fishing harbor route from Mangaluru to Thiruvananthapuram was 'Protect Sea, Coast, Inland Water Bodies and Fish Resources'. The value of interlinked water bodies is better understood now than ever before.

Bittu Sehgal, member the Coastal Task Force and founder of India's premier wildlife magazine, *Sanctuary*, has indicated that

> *land developers around the country are working overtime with state government officials to encash literally thousands of crores of rupees worth of public lands. There is a similar move afoot for forestslands and together the attack on coastal belts and forest belts will alter the survival ecology of over 300 million*

people within the next three to five years. To my mind this represents a more serious threat than any other development project or anti-people initiatives ever initiated on the Indian subcontinent in the past 50 years. (1998, in correspondence with Mr. Vishvanath Anand, Addl. Secretary at the Ministry of Environment and Forests)

Inconsistent Judicial Interpretation of Land-Use Restrictions

Fragile ecosystems of water bodies have not been appropriately located in many judgments delivered in the last two decades. The approach to CRZ law has been inconsistent and has led to incomplete solutions to the problem of fragility in riverbeds and along coasts.[11] Whenever these contestations are pulled into the courtroom there are claims and counterclaims over the nature of ecosystems, state responsibility and the rights of local people. There exists enormous disparity in legal claims presented before the judges in interpreting both the 'resources' and 'public welfare' on different occasions.

Inconsistent judicial interpretations can be demonstrated to revolve around three pillars: the CRZ land-use restrictions, public trust doctrine, and the irresponsible and unaccountable governance institutions:

- In the CRZ land-use restrictions judges have sought various definitions for defining coasts and land-use diversions. At times courts have diverted the onus of decision making to the CZMA[12] and in some cases have allowed CRZ violation.[13]On the issue of land reclamation the High Court of Kerala did not find any serious impairment to aquatic resources, ecology and environment in the land reclamation demanded by the Goshree Project,[14] and even declared that the land proposed does not come under the CRZ area. The judiciary has interpreted the land reclamation in different ways as well. In the case of the Institute of Social Welfare vs. State of Kerala (1996 (1) KLT 718 = AIR 1997 Ker.45) only such land reclamation as would disturb the natural course of seawater was prohibited and the court even insisted that any neglect of the aforementioned interpretation may lead to unregulated constructions of various kinds. However, two years later the court applied a relatively liberal interpretation to the above decision and allowed the Goshree Project to reclaim 25 hectares of land near Marine Drive. This led the Courts to incline toward soft decisions and dismiss clear action, as was initiated by Justice A V Ramakrishna Pillai[15] when he ordered the demolition of buildings which encroached upon backwaters. Moreover, it was only after the country's top accountability body CAG exposed the defiance of CRZ regulations by builders that government bodies and the Courts sprang into action. There is a need for revisiting regulations which regulate land use around coastal and catchment zones. The fundamental guiding principle of coastal conservation is best expressed by Justice A V Ramakrishna Pillai, who ordered demolition of DLF construction around the Chilavannoor backwaters in Ernakulam: 'Nature, which is the property of the nation, cannot be allowed to be scrambled by a minority violating all laws'.[16] Ironically, the division bench of the Kerala High Court comprising Justice Thottathil B. Radhakrishnan and Justice Babu Mathew P. Joseph passed the order granting a three-month stay on an appeal filed by the DLF[17] against the single bench order and granted a three-month stay to demolish the DLF apartment complex for violating the CRZ notifications. As CRZ law applies equally to the riverbed and catchment zones, this discussion could include the battle over the Yamuna riverbed for the construction of the Commonwealth Games Village in 2010. The controversy over whether or not the construction was actually on a riverbed became an argument for allowing the construction. As already mentioned, while a number of judgments treat riverbeds as a waterbody which cannot be used for construction, a Supreme Court Bench headed by the Chief Justice himself declared the riverbed as 'land' and therefore open for construction.

- The doctrine of public trust is a symbolic but logically explainable understanding between the state and citizens under natural law. The Hon'ble Supreme Court in M.C. Mehta v. Kamal Nath and Others (1997, SCC 388) articulated the doctrine of 'public trust', the theoretical thrust of which was elaborated as 'certain common properties such as rivers, seashores, forests and the air are held by the Government in trusteeship for their free and unimpeded use of general public. Following the judgement in the above case, the State and High Court shall endeavor to protect and improve environment'.[18] Much of this law is guided through Article 48A and Article 51A inserted in Part IV of the Constitution through the 42nd Constitutional Amendment Act 1976. While the former places a responsibility on the state, the latter emphasizes citizens' duty to protect the environment. An important but terse message which emerges from the doctrine of public trust is to reject and condemn private ownership of environmental resources as complete defiance of natural justice. This suggests that forests, water bodies and air belong to people and the state is merely a custodian of this 'people's property' as a trustee. Writing on the relationship of public trust doctrine and environmental conservation, Sax[19] suggests that '[o]ur contemporary concerns about 'the environment' bear a very close conceptual relationship to this venerable legal doctrine (of public trust)' (1970: 164). It is primarily due to the doctrine of public trust that courts tend to recognize the ***doctrine of legitimate expectation***[20] whereby the government is expected to demonstrate some form of regularity, predictability and an assurance rather than a drastic turn-around from an expected procedural and substantive system of functioning. It is an unwritten code of standard behavior which the citizen expects from the state.

- Governance of CRZ has been tarnished due to the presence of multiple administrative authorities and each being 'not responsible' or waiting for another department to take a stand on an issue. There are many decision makers on CRZ such as the Ministry of Environment, Forest and Climate Change (UNEFCC), the State Coastal Zone Management Authority (KCZMA), the State-Level Environment Impact Assessment Authority (SEIAA), the State Expert Appraisal Committee (SEAC), the Port Trust (PT), the Fisheries Department (KFD) and the Municipal Corporation. On November 3, 2011, in exercise of the powers conferred by sub-section (3) of section 3 of the Environment (Protection) Act, 1986 (29 of 1986) and in pursuance of the Government of India in the Ministry of Environment, Forest and Climate Change notification number S.O. 1533 (E), dated September 14, 2006, the Central Government constituted SEIAA and SEAC in Kerala. The overlapping jurisdictions of Central and state government on one hand and of hierarchies of state agencies and departments on the other, inappropriate standardization of jurisdictional boundaries and a loose time frame without clear penal action have dispersed the norms of accountability within the whole state's administrative framework. Justice Ramakrishna Pillai's demolition orders on DLF encroachments on Chilavannoor Road, Ernakulam backwaters, highlights the administrative quagmire as well as their lethargy. The builders found ample negotiating space to obtain all clearances for construction in the CRZ. The Town Planning Standing Committee made the city corporation issue permits, a basic requirement to obtain an electricity connection, but even where these permits, approvals and occupancy certificates were not given, all buildings, even the 7- and 20-story apartments, were found to be lit with electricity and had working elevators.

The Three Major Deficits of Governance: Study of the Kerala Backwaters Case 2014

First: *Accountability for granting clearances was not fixed*
The process for approval is tough, and begins with the Municipal Corporation allocating land for construction. The KCZMA then approves the plan through environmental clearances from the Ministry of Environment, Forest and Climate Change. The departments of the District Development Commissioners and District Planning Officers are part of the urban local body which gives environmental clearances to projects in CRZ areas.

Second: *Monitoring and Reporting agencies have no power to stop non-compliant projects immediately*
The urban local body as part of District Administration and Town and Country Planning failed to enforce land-use and municipal laws, building structural plans and floor size regulations. The Vigilance and Anti-corruption Bureau (VACB) made surprise raids at the construction sites of ten municipalities only after CAG reported violations in its report. When the Corporation sent notices to some of them it was more a ritual administrative exercise to be on the right side of the law rather than any commitment to stop illegal construction.

Third: *Lack of respect for internal procedures and EIA.*
In most of the violations it was found that the Municipal Corporation gave permission for construction without addressing and insisting on the need for obtaining even the required mandatory approvals from the CZMA and the Central Ministry (MOEFCC). CZMA later informed

the corporation about the ongoing construction of 13 buildings but the corporation sent notices to only one construction company to stop work. The role of the corporation secretary also vacillated between different options. The Electricity Board overlooked the mandatory requirement of possessing Occupancy Certificates from the Municipal Corporation for obtaining electricity connections for the buildings.

For coastal and catchment zone protection there are three major statutory bodies managing encroachments: the CZMA, the SEIAA and the SEAC. Their relationship with one another and their coordination with other statutory bodies, that is, the District Town and Country Planning Department, the State Industrial Corporation, the Urban Development Authority and most of all the local Panchayat Authorities, have become increasingly unaccountable. This sometimes pushes fragile regions into greater vulnerability. A recent trans-disciplinary study (Auerbach et al. 2015: 1–5) by scholars from the social sciences and earth sciences has revealed that much of what administrators do in preventing floods and promoting environmental conservation actually exacerbates these conditions and has even led to sea level rise and flooding.

Addressing the Overload of CZMA

CZMA was created by the Central Government under the powers conferred through sub-sections (1) and (3) of section 3 of the Environment (Protection) Act 1986. Its mandate stretches from livelihood security to disaster risk reduction policies:

> *... in exercise of the powers conferred by sub-section (1) and clause (v) of sub-section (2) of section 3 of the Environment (Protection) Act,1986 (29 of 1986), the Central Government, with a view to ensure* ***livelihood security*** *to the fisher communities and other local communities, living in the coastal areas, to* ***conserve and protect coastal stretches****, its* ***unique environment and its marine area*** *and to promote* ***development through sustainable manner*** *based on scientific principles taking into account the dangers of* ***natural hazards*** *in the coastal areas,* ***sea level rise due to global warming****, does hereby, declare the coastal stretches of the country and the water area upto its territorial water limit, excluding the islands of Andaman and Nicobar and Lakshadweep and the marine areas surrounding these islands upto its territorial limit, as Coastal Regulation Zone (hereinafter referred to as the CRZ) and restricts the setting*

up and expansion of any industry, operations or processes and manufacture or handling or storage or disposal of hazardous substances as specified in the Hazardous Substances. (Handling, Management and Transboundary Movement) Rules, 2009 in the aforesaid CRZ...

Principal Secretaries and Secretaries from the key state departments such as science and technology, revenue, local self-government, fisheries, environment and industries are appointed as members for a tenure of three years. Members are not 'just anyone' representing that particular department of the state government, but are selected because they are heading these departments, which makes all the difference in CRZ implementation. By virtue of being the head of their own department the additional charge of CZMA brings an overload of additional responsibilities. These additional and also secondary assignments are often treated as avoidable, aside from attending meetings called by the Member Secretary from time to time. Most Member Secretaries complain about the difficulties of getting everyone together and also getting everyone to rise above narrow departmental boundaries, and about uncompleted homework expected from each of them on a case-by-case basis mentioned in the agenda. Most of these Secretaries are transferred much before completing the tenure of three years, and the next incumbent is not completely aware of the sensitivity of most projects outside the regular departmental responsibilities. Moreover, the unspoken tension which rips apart any committed administrative action is the lack of attention given by the new incumbent to the decisions and projects attended to by the previous one.

Conclusion

The law of conservation is not inherently weak. However, what makes it weak is the combination of the failure of state leadership, administrative and judicial corruption, and lack of ecological awareness. There are many factors which prevent sensitization, resonance and rigor in environmental governance. The most important one is the need for public feedback (and the application of the *audi alteram partem rule to expect all parties to be heard appropriately before taking a decision)* and a regular supervisory and evaluation committee which can best be arranged through the central ministry (MEFCC). In the Case of protecting CRZ of Kerala Backwaters, Justice A.V. Ramakrishna Pillai observed,

> *The purpose of these laws is to preserve nature for posterity. If the violation of the laws is allowed to become the order of the day, the existence of life would be at peril. Right to life guaranteed by the Constitution takes innumerable rights, including the right to enjoy nature in the present form. Indiscriminate invasion of nature to the detriment of others is an invasion of right to life.*

All this can be achieved if public institutions perform as custodians of the trust property rather than agents of its destruction and a cause for disaster.

Notes

1. In the order dated May 3, 2005 in W.P. (C) No. 689/2004; order dated March 31, 2003 in W.P. (C) No. 8227/2002.
2. W.P. (C) No. 6729 & 7506 of 2007.
3. W.P. (C) No.689/2004 Order dt. May 3, 2005, W.P. (C) No. 8227/2002 Order dt. March 31, 2003.
4. 'Deltas Sinking and Shrinking as Dams Curb Steady Flow of Fresh Water to the Coasts', *The New Indian Express*, January 28, 2015, 09:09 pm, Study of Coastal Zones National Centre for Sustainable Coastal Management, 'India has 7500 km coastline (~5400 km on the mainland) and about 250 million people live within 50 km distance from the shore. India's mainland coastal stretch of 5422 km have undergone tremendous changes due to varying natural and human induced coastal activities'. (http://ncscm.org/cms/geo/pdf/research/High%20resolution%20Erosion.pdf)
5. 'Indian Coastline Rapidly Eroding', *Deccan Chronicle*, Chennai January 28, 2015, Shoreline Change Assessment for Kerala Coast available at http://ncscm.org/cms/more/pdf/ncscm-publications/kerala_fact_sheet.pdf and Purvaja and Senthil (2011) available at http://www.ncscm.org/reports.php
6. The line on the land up to which the highest waterline reaches during the spring tide.
7. Inserted by S.O 550 (E) dt. May 21, 2002, vide *Gazette of India* (Extra) No.470.
8. Indian Council of Enviro-Legal Action vs. Union of India & Others 1996(4) JT SC 263
9. Ratheesh and Others Vs. State of Kerala and Others, W.P. (C) No. 19564/11 decided on July 25, 2013.
10. Union Territory of Lakshadweep vs. Seashells Beach Resort. AIR 2012 SC 2309:2012(6)SCC136:2012 AIRSCW3343., Union of India vs. Chennai

Metropolitan Development Authority, 2006(4) KLT.SN.117.P.84:2006(4) CTC460. J., Piedade Filomena Gonsives vs. State of Goa. AIR 2004 SC 3112: (2004) 3 SCC 445:2004 AIRSCW 2302.

11. Huffman, 'In a sense, the widespread misrepresentation of the history of the public trust doctrine is apt because the lawmakers themselves often have been party to the distortions' (2008:8).
12. P.A. Fazal Gafoor vs. State of Kerala. Special Leave to Appeal (Civil) No. 5038/2002. April 25, 2003.
13. Goa Foundation vs. Diksha Holdings Pvt. Ltd.2001 (2) SCC 97=AIR 2001 SC 184.
14. Jacob Vadakkancherry vs. State of Kerala AIR 1998 Ker. 114.FB.
15. Antony A.V. Vs. Corporation of Cochin WP(C) No. 27248 of 2012 (E), Manu/KE/20141/2014 decided on December 8, 2014 in the High Court of Kerala.
16. Manupatra, Manu/Ke/2041/2014. p. 9.
17. *Deccan Chronicle*, '3 Month Stay on Order to Demolish DLF Flat', December 21, 2014, 05.12 am IST.
18. Vishnu Motor vs. Vishakhapatnam Urban Development Authority, 2002 (4) ALT 746.
19. Sax, Joseph L. (1970) The Public Trust Doctrine in Natural Resource Law, Effective Judicial Intervention, *Michigan Law Review*, Vol. 68:471, January.
20. In Sethi Auto Service Station and Another vs. Delhi Development Authority & Others (2009) 1 SCC 18. the court observed that this doctrine of legitimate expectation was at the root of the constitutional principle of the Rule of Law.

References

Auerbach, L. W., Goodbred Jr., S. L., et al. (2015, January 5). Flood Risk of Natural and Embanked Landscapes on the Ganges–Brahmaputra Tidal Delta Plain. *Nature Climate Change*. 1–5. Published online.

Costanza, R., d'Arge, R., de Groot, R., Farber, S., Grasso, M., Hannon, B., Limburg, K., Naeem, S., O'Neill, R., Paruelo, J., Raskin, R. G., Sutton, P., & van den Belt, M. (1997, May 15). The Value of the World's Ecosystem Services and Natural Capital. *Nature, 387*, 253–260.

Dearing, J., Yang, X., Dong, X., Zhang, E., Chen, X., Langdon, P. G., Zhang, K., Zhang, W., & Dawson, T. P. (2012). Extending the Timescale and Range of Ecosystem Services Through Paleoenvironmental Analyses: The Example of the Lower Yangtze Basin. *Proceedings of the National Academy of Sciences, 109*, E1111–E1120.

Huffman, J. (2008). A History of the Public Trust Doctrine. *Duke Environmental Law and Policy, 18*, 1.

Purvaja, Ramesh., & Senthil Vel, A. (2011). *National Assessment of Shoreline Change: Odisha Coast*. NCSCM/MoEF Report 2011–01, 57 p. Available at http://www.ncscm.org/reports.php

Sax, J. L. (1970). The Public Trust Doctrine in Natural Resource Law, Effective Judicial Intervention. *Michigan Law Review, 68*(January), 471.

Amita Singh is Professor of administrative reforms at the Centre for the Study of Law and Governance, JNU, New Delhi, and founder chairperson and transdisciplinary scholar at the Special Centre for Disaster Research, JNU, New Delhi.

Environmental Sociology of Floods in the Colombo District of Sri Lanka

Dinushika M. Yapa Abeywardhana

Introduction

Environment means everything around us and the interaction between environment and society is being studied in environmental sociology. Disasters are natural as well as man-made phenomena that can be considered among those environmental problems. According to Drabek (2005), disaster is a situation that causes a certain population to face hardships and damages property, earnings and resources beyond recovery. Flooding is prominent and there are persuasive evidences that the costs of extreme weather events, with flooding as a major contributor, have been exhibiting a significant upward trend (UNISDR 2011). Therefore, a holistic perspective is needed to understand the changing rainfall-driven flood risk as well as the man-made flood risk and to reduce adverse impacts caused by them. According to reported world disaster statistics within the period from 1900 to 2006, 30% of disaster events are flood events, and from the

D. M. Yapa Abeywardhana (✉)
Department of Sociology, University of Colombo, Colombo, Sri Lanka

A. Singh et al. (eds.), *Development in Coastal Zones and Disaster Management*, Disaster Research and Management Series on the Global South, https://doi.org/10.1007/978-981-15-4294-7_18

beginning of ninetieth century, the frequency of flooding has increased particularly in the Asian Region. Moreover, Sri Lanka's second critical disaster is flooding and the first critical disaster is drought, according to present statistics (Liyanarachchi 2017: 25).

Consequently, the present study inquired about the flooding incident of May 2016, which had an adverse impact on the urban sector of Sri Lanka. Unexpected disasters may not only destruct people's day-to-day routine and their physical assets but also can affect both short-term and long-term physical and psychological lives of victimized people (Abeykoon and Samarakoon 2017). The graph below shows the number of people who were affected by flooding within the past few years in Sri Lanka (Chart 1).

There are few things that can be identified through the above chart. The tendency of increase in flooding can be seen within the period from 1982 to 2017. Apart from that, the number of people who are affected by flood has fluctuated and it seems like the increasing level is multiplied by four levels in an outbalanced manner. It further depicts the number of

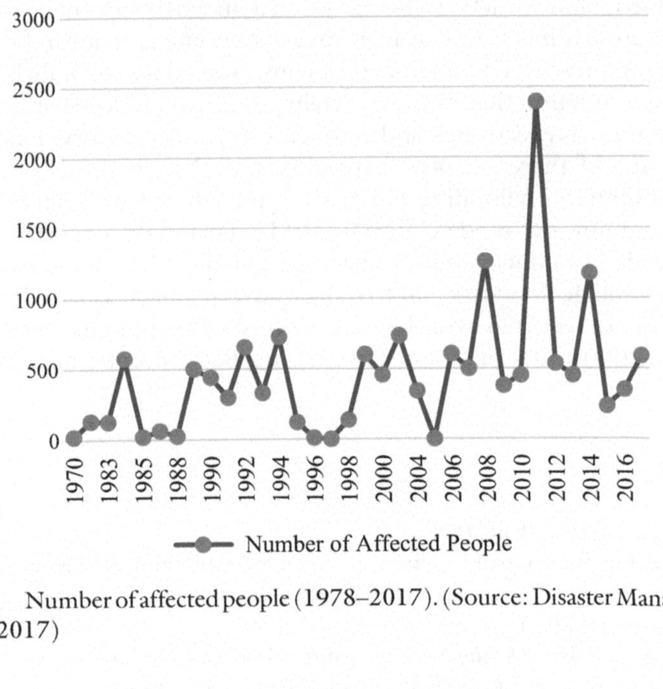

Chart 1 Number of affected people (1978–2017). (Source: Disaster Management Centre 2017)

affected people in 2003, 2008 and 2007. The highest level of affected people can be identified in 2007 and that level and its uneven fluctuation further warns about future flood risk. In this background, the importance of disaster management has emerged and through that it is expected to increase the capacity of disaster resilience of the whole country, including individually affected people (Dhanapala 2012: 31).

Since flooding occupies a primary place among natural disasters in Sri Lanka, there is a long history with regard to its management. It means Sri Lankans have 200 years of local practices which was used to flooding disaster management. In the period of British and Portuguese colonization, the colonizer and colonized have tried to practice structural mechanisms for flooding disaster management in the Kelani, Nilwala and Gin Rivers. After the redemption of British colonization, the government has introduced a few projects to reduce flooding disasters in the wet zone of Sri Lanka. For instance, in 1960 a foreign organization in collaboration with the Department of Irrigation in Sri Lanka started a project named, "Three Basis Project" to mitigate floods in Kelani and Nilwala river. Another project proposal was introduced in 1961 by Technopromeyport Organization of Russia with Department of Irrigation to mitigate Kelani river floods. But the issue was those projects were stagnated at the basic levels and the impacts of flooding disaster was highly increased (Liyanarachchi 2017: 115). Therefore, the increase in the frequency of unpredictable floods led to a focus on modern flood risk management methods such as Integrated Flood Management (IFM), Flood risk assessment and so on. While these structural mechanisms are considerably focused on this upcoming issue, they also need to focus on social factors with the intention of reducing man-made factors that can impact on the frequency of floods in urban sector. Apart from that, it is important to create a social framework to minimize disaster-related issues in the context of flooding.

This chapter explores the impact of flooding on urban people's life and presents resolutions to reduce its harmful impacts. Since flooding can destruct human everyday life as well as their whole structural life components, it is important to address this upcoming issue. As Park (1992) states, urban life is considerably complex and problematic; its impact and adverseness towards lives should be carefully reviewed (Niriella 2014). Furthermore, the adaptability and coping with the situation creates different set of practices amongst the said community. This emphasizes the importance of a sociological reading in the context of flooding disaster, particularly in a developing country in the South Asian Region. This study has several objectives: (1) Identify man-made factors affecting flooding, (2) Understand people's experience to vulnerability of flooding, (3) Identify the impact of flooding on people's day-to-day lives.

Research Methods

This is an explorative study and uses both quantitative and qualitative data from the research field of Kohilawatta, Kolonnawa, in Colombo district, Sri Lanka. Purposive sampling method was used to select households. The author administered the questionnaire which contains pre-coded, open-ended questions. Semi-structured and in-depth interviews with observation were used to collect qualitative data. A questionnaire was used in order to get a basic idea about the demographic and socio-cultural factors of the respective field. Consequently, forty households were selected for the questionnaire through the purposive sampling method. Ten semi-structured interviews were administered with the respondents who were selected from questionnaires and four in-depth interviews were administered with key informants selected through purposive sampling method. Non-participatory observation was also utilized to a certain extent. Collected data were analysed using the thematic content analysis method. Apart from the above primary data-collection methods, secondary data from the literature of respective disciplines were also used.

Results/Discussion

Flooding and Culture of Vulnerability

According to Manawadu and Liyanage (2008), population density of the Colombo district has increased at the given rate: 1981 – 0.8%, 1911 – 2.7%, 1971 – 4.2%, 2001 – 1.9%; this fluctuation shows an uneven pattern of urbanization in Sri Lanka. However, people select Colombo, the main commercial city of Sri Lanka, as they can easily access most of the urban facilities. This creates a considerable demand and high pressure on the lands.

> *People try to buy cheap vulnerable lands in this area. Because they just want to live in Colombo. They try to skip their vulnerability towards flooding. But this create big problems that cannot be ignored and their lives are always in risk because of their decision. (A respondent, Field Data, 2017)*

It emphasizes various man-made factors which are related to urban flooding. Among the main findings of the study, urbanization impact on the increasing population density in the respective area was prominent. People were attracted to this area because of social, environmental as well as financial reasons. Affordability, facility accessibility, river-based livelihoods are a few of them. According to reports of Grama Niladhari (GN) officer (2015), the majority of the people in the field come from low-income backgrounds and they cannot afford expensive, less vulnerable lands near the city. Since they want to live near Colombo because of the above-mentioned reasons, they choose to live in an environmentally sensitive, flood-prone area, without considering any side-effects of that decision.

According to Strydom (2002), there is a historical relationship between vulnerability, environment and society. Beck (1992) states that the society overflows with vulnerabilities and vulnerability can be identified as a development theory in the present post-modern society. Douglas and Wildavsky (1982) further explains the vulnerability while examining its characteristics; she states that, vulnerability is defined on the basis of cultural structures which are created under people's interpretation. Therefore, the said data also proves this interpretational state and people have neglected their vulnerability towards flooding while giving priority to their complex social needs.

Environmental Consciousness Leads Mass Consumption

According to the resource profile of Kolonnawa divisional secretariat (2016), the environmental and geographical structures of the respective area are not suitable for settlements and those areas are wetlands since long time ago (Table 1).

Table 1 Awareness about environmental issues in the country

Response	*Count*	*%*
Yes	15	37.5
No	25	62.5
Total	40	100

Source: Field Data, 2017

Since the researcher needed to understand the environmental consciousness of people, the respondents were questioned about people's awareness about environmental issues in the country. The above data proves that they have less environmental consciousness as well as task-oriented attitudes. The percentage of respondents who are not aware about present environmental issues is significantly higher (62.5%) than those who are aware (37.5%). This is further proved through observed data. People tend to fill those wet lands, including paddy fields and small water slumps, and start building their own houses on the lands without considering imminent issues and suffering after disaster. As deduced from most of the interviews, there was not much legal interference to those activities as political patron–client relationship is strongly influential within the field and even politicians contribute to the continuation of illegal constructions, including personal housing and commercial buildings.

> *We always need to build our house better. We have faced flooding for years. But once we finish our house with few levels, we can face it without any problem. Government and NGOs help us with money when flooding came. Now we have enough money to do build our houses better than others. (A respondent, Field Data, 2017)*

Apart from previously mentioned factors, human activities which are based on symbolic consumption create less resilient land structure to flooding. According to Baudrillard (1989), the consumption builds inflexibly with the collective social competitiveness. When it is substituted to the present study, above semi-structured interview data show that people build or renovate their houses particularly using flood compensations to symbolize their social capacity as compared to others, without considering the pressure to the lands. Doing this further creates less resilient land structures, as mentioned in the above discussion.

According to Project Insights (2013), floods are caused by a combination of factors, such as unauthorized constructions (slums and shanties) obstructing the water flow, clogging and lack of regular maintenance of drainage canals, backwater build-up in the main canal system and commercial development in wetland reservations (Fernando 2017: 1027). Therefore, it is evident that these human activities are thoroughly combined with the emergence of sudden flash floods than rainfall-driven floods.

The side-effects of irresponsible environmental behaviours come back to people as a destructive disaster, named floods. Even though the disaster outwardly seems only like a destructive event, the above data revealed that people experience it in various ways. In other words, since they have adapted to this annual weather-related disaster, they do not see it as an abnormal issue. This point is further confirmed through the table below.

Environmental Adaptability and Coping Strategies

When someone frequently deals with hard or complex situations, they may adapt or cope with that particular incident. This may be practical with regard to the community living in the endangered area. The graph below displays people's own views about their experience of facing flooding events (Table 2).

As the graph explained, the highest percentage (42%) shows that flooding is a disturbance to people's life. However, when all other percentages of other legends get together, it is higher (45%) than this percentage. Therefore, this study indicated that people have normalized their vulnerability towards flooding and have adapted to it. However, this adaptability and overestimated coping capacity have led those people's life towards more risks as present floods are highly unpredictable. The most recent example for that is the flooding incident of May 2016, which is discussed in this study.

Negative and Positive Impacts of Flooding Events

In the flooding event of May 2016, 340,150 million people were affected, 503 houses were totally destroyed and 3793 houses were partially destroyed (Disaster Management Centre 2017). According to the field data (2017), all houses of the respective field were also affected and half of

Table 2 People's view on the impact of flooding to their life

Legend	*Count*	%
Disturbance to the day-to-day life	17	42
Risk to life	3	7.5
Fear of living the area	2	5
Not a big issue	11	27.5
Normal situation	7	17.5
Total	40	100

Source: Field Data, 2017

the community was displaced due to not paying heed to pre-disaster warnings and staying at home until water level rose up. People had to face various hardships as they lost their livelihoods (43.5%), hindrances to the education of children (25%), increasing health issues (19%) and destruction of cultivated lands (12.5%). Since it was the largest flooding event after the 1989 flooding event of Sri Lanka, people were significantly affected in the post-disaster period. Moreover, this destructive event caused unprecedented loss as proven through the data, where people suffered various psychological impacts such as PTSD, psycho-social depression and so on.

The field has a mixed ethnic population. Among them, 74% are Buddhists, 22% are Muslims and the rest are Hindus and Catholics. These ethnic differences were not particularly considered during the disaster. According to the majority (8) of interviews, since every person of the respective field was affected by this disaster, each and every ethnic group came together to help each other, without caring about any religious or ethnic differences. Therefore, social cohesion has been developing within the disaster context. On the other hand, religious priests have played a prominent role in this context. They have been leading the mobilization of people, which highlights the importance of helping one another in this crucial situation.

When considering those points, it is important to present a better resolution to get rid of such malfunctions of disasters like flooding. The study proposed that a better environmental behaviour should be promoted among people through ground-level awareness programmes and community-based activities. Since most of the people of the respective fields have a weak financial background, they must be educated to live sustainably in environmentally sensitive areas because it will not be a practical solution to resettle them in another area as their identity as well as physical stability are attached with their residences. If an environmental-friendly housing system is introduced, these problems may be reduced.

Moreover, proper implementation of pragmatic conceptualizations like 'Riparian Community' within its capacity may reduce the flooding frequency and increase the resilient capacity. Therefore, to reduce the flood risk as well as its adverse impacts, environmental-friendly techniques should be developed; to fulfil this, holistic approaches are needed and people must be educated to be more environmentally responsible.

Conclusions

Flooding dominates the "upcoming disaster" range. Human activities are based on political economy, creating neither more vulnerable nor weak environment structure that can be easily affected to the flooding. On the other hand, those less conscious environmental actions indirectly contribute to the increasing of unpredictable floods each year. These situations can be particularly identified through the urban context in Sri Lanka as when compared to the rural sector, urban sector comprises of high complex social, environmental, political and economic backgrounds. Those inter-related factors strongly combine with disasters, unsustainability and instability of the social system.

People's culture of vulnerability manipulates their social behaviour and increases the risk of their social lives as well as physical well-being. Therefore, this culture of adaptation generates both functions and malfunctions. In spite of that, the symbolic consumption characterizes various aspects of human needs and further contributes to the increasing of flooding. On the other hand, the collective consciousness has created through the heterogeneity in the respective field. A disaster was able to bring together different ethnic groups separated by their culture or beliefs. Therefore, this shows that heterogeneity helps to create homogeneity in the context of disaster.

The nature of power relations worsens this background and further disturbs the creation of sustainable environment. Proper legitimation is needed to address those politics-related issues as they carry on so much of power on people. Furthermore, a discourse must be created on disaster-related issues which would be helpful to carry out practical and above-mentioned resolutions to reduce flood risk in the Sri Lankan urban sector.

References

Abeykoon, W., & Samarakoon, U. (2017). *Impact of Cultural Education on Social Cohesion After a Sudden Disaster: With Reference to Two Flood Affected Locations in Colombo District-Sri Lanka*. Nagoya: The International Academic Forum. Available from: https://papers.iafor.org/3579. Accessed on 29 Oct 2017.

Baudrillard, J. (1989). *Simulacra and Simulation*. Ann Arbor: University of Michigan Press. Available from: http://www.e-reading.club_3486. Accessed on 3 Jan 2018.

Beck, U. (1992). *Risk Society*. London: Sage.

Dhanapala, H. A. (2012). *Disaster Management*. Nugegoda: Sarasavi Publication.

Disaster Management Centre. (2017). *Hazard Profile of Sri Lanka*. Colombo: DMC.
Douglas, M., & Wildavsky, A. (1982). *Risk and Culture: An Essays on the Selection of Technological and Environmental Dangers*. Berkeley: University of California Press. Available from: http://www.jstor.org/stable/10.1525. Accessed on 15 Nov 2017.
Drabek, T. E. (2005). *Predicting Disaster Response Effectiveness*. International Journal of Mass Emergencies and Disasters. Newark: University of Delaware. Available from: http://socialscience.focusonflooding.org. drabek213. Accessed on 5 Aug 2017.
Fernando, N. (2017). *Voluntary or Involuntary Relocation of Underserved Settlers in the city of Colombo as a Flood Risk Reduction Strategy: A Case Study of Three Relocation Projects*. Bangkok: Procedia Engineering. Elsevier.
Liyanarachchi, P. (2017). *Flooding Disaster in Sri Lanka: Process, Impact and Management*. Angoda: Author's Publication.
Manawadu, L., & Liyanage, M. (2008). *Third World Urbanization*. Borella: Wijesuriya Publucations.
Niriella, C. N. (2014). *Urban Sociology: Community, Neighborhood, Class Distribution*. Nugegoda: Sarasavi Publication.
Strydom, P. (2002). *Risk, Environment and Society Ongoing Debates: Current Issues and Future Prospects*. Philadelphia: Open University Press.
UNISDR (United Nations International Strategy for Disaster Reduction). (2011). *Revealing Risk, Redefining Development*. Geneva: UNISDR.

Dinushika M. Yapa Abeywardhana is an assistant lecturer in Sociology attached to the Department of Sociology, University of Colombo, Sri Lanka. She has been awarded Gold Medals for the Best Student of Social Theory and Development Sociology at the General Convocation of the University of Colombo, 2018. Currently, she is studying for Masters of Arts in Sociology at the Department of Sociology, University of Colombo, specializing in Environmental Sociology. Her research interests are Sociology of Disaster, Environmental Sociology, Animal and Society, Social Theory, Development Sociology and Cultural Anthropology.

Loss and Damages from Cyclone: A Case Study from Odisha, a Coastal State

Trupti Mishra and Krishna Malakar

Introduction

Cyclones are a significant risk to lives and property in coastal areas, and cause severe loss and damages to communities. In 2018, cyclones resulted in a loss of 56 billion USD globally (Munich RE 2019). India (which is exposed to almost 10% of the world's cyclones), and particularly its eastern coast, is one of the most vulnerable countries (Government of India 2019). In the last two decades, from 1998–2017, India has lost around 80 billion USD due to disasters and ranks among the top ten countries in terms of absolute losses suffered (Wallemacq and House 2018). Among many cyclones, Phailin (2013) was one of the most disastrous in the recent past in India. It made landfall in Gopalpur in the state of Odisha (IMD

T. Mishra (✉)
Shailesh J. Mehta School of Management, Indian Institute of Technology Bombay, Mumbai, India
e-mail: truptimishra@iitb.ac.in

K. Malakar
Indian Institute of Technology Bombay, Mumbai, India
e-mail: krishna.malakar@iitb.ac.in

A. Singh et al. (eds.), *Development in Coastal Zones and Disaster Management*, Disaster Research and Management Series on the Global South, https://doi.org/10.1007/978-981-15-4294-7_19

2014), and resulted in a total loss of 1450 million USD to the state (Government of Odisha 2013). This included loss of and damages to housing, public buildings, roads, urban and rural infrastructure, agriculture, horticulture, irrigation, livestock, livelihood (fisheries, Micro, Small and Medium Enterprises and others), energy/power, forest and plantations. However, the United Nations applauded the efforts of the state as it was well-prepared for the cyclone and evacuated around 1 million people from the coast, which resulted in low fatalities (Senapati 2013).

The marine fishing community who reside in coastal areas is one of the most vulnerable to cyclones. In 2013, when Cyclone Phailin devastated many parts of Odisha, the fishing community was one of the most affected (Bal 2018; Government of Odisha 2013; *The Hindu* 2013). Around 11,000 marine fishermen suffered damages from the cyclone, and the losses of the fisheries sector amounted to 97.53 million USD (Government of Odisha 2013).

The Rapid Damage and Needs Assessment Report (regarding Phailin) provides a comprehensive assessment of the loss and damages suffered by various sectors in Odisha from the cyclone (Government of Odisha 2013). It provides a state-level estimation of the total damages. However, community-specific studies on loss and damages are also needed to gain an understanding of the impacts of cyclones regarding aspects other than assets, such as time needed to restart livelihood and build houses. This is particularly important in the case of communities having specialized natural resource-dependent livelihoods such as fishing. Thus, this chapter focuses on the vulnerable fishing community and attempts to document both economic and non-economic loss and damages. It studies the losses in livelihood, housing and overall household well-being, for example, occurrence of injury or illness. It also gives an overview of the time and cost for the community to return to their livelihoods and rebuild their houses. Thus, it offers greater detail on the damages, which are beyond monetary losses, incurred by the community. Further, loss in terms of time has also not been assessed in reports on the cyclone. Assessments of loss and damage can help in planning and directing financial investments in vulnerable communities/sectors/regions. Such studies have implications for policy on adaptation and disaster management. Understanding loss and damages can prepare communities for the future by pushing them to design interventions to improve their adaptive capacity and minimize losses (van der Geest and Warner 2015).

Existing Studies on Loss and Damages from Cyclones

The literature on loss and damages from extreme weather events includes country-level studies on estimating economic losses from events such as cyclones and floods (Crompton and McAneney 2008; Pielke and Landsea 1998; Zhang et al. 2009). These studies are mostly from developed countries such as the USA and Australia and data-rich countries such as China. Pielke and Landsea (1998) normalized decadal monetary losses due to hurricanes in the USA. The study suggested that changes in population and possession of assets/wealth also need to be considered while analyzing increase or decrease in losses over time. Crompton and McAneney (2008) assessed changes in normalized insured losses from meteorological hazards in Australia over 40 years. The paper finds that the rise in number and value of housing/dwellings resulted in increasing losses. Similarly, Zhang et al. (2009) have analyzed damages due to tropical cyclones in China (from 1983 to 2006). They found that the increase in losses was mostly because of China's economic development.

Further, there have been studies from developing countries on estimating the loss and damages to regional communities from disasters such as floods (Arunodhaya et al. 2016; Kakakhel 2015; Patankar and Patwardhan 2016). These studies are based on primary data collected from households affected by these events. Kakakhel (2015) gauges the loss and damages in a flood-affected village in Pakistan. The study attempts to quantify both economic losses such as income and assets, and non-economic losses such as health and public infrastructure. Arunodhaya et al. (2016) undertook a detailed survey to estimate the losses to homes, goods, occupational tools and effect on working days of the population impacted by floods in the city of Chennai (India) in 2015. Patankar and Patwardhan (2016) quantified uninsured losses suffered by households and small businesses in Mumbai due to floods in 2005. Rabbani et al. (2013) focused on loss in rice production due to salt intrusion as a result of Cyclone Aila in Bangladesh. Further, Warner and Geest (2013) did not directly quantify the losses but gave an overview of the stresses resulting in loss and damages to population in nine vulnerable and developing countries. They also showed that often economic loss and damages are recorded more in the literature than non-monetary losses such as health and livelihood (van der Geest and Warner 2015).

In contrast to studies based on primary surveys, Dutta et al. (2003) estimated loss and damages in a river basin in Japan using simulated flood parameters and mathematically combining a hydrologic and loss estimation model.

Thus, there are country-level, regional-level and simulated studies on quantifying loss and damages. There are also qualitative studies which describe loss and damages to communities from disasters. For example, one of the reports on Cyclone Aila in Bangladesh (ReliefWeb 2009) lists the number of damaged households, livestock and fishing gear and describes the experiences of few affected individuals. The present study is a regional-level assessment focusing on a vulnerable community residing along coasts and involved in marine fishing. Family, housing and livelihood gear of these communities are prone to loss and damages. This study adds to the limited literature by examining loss and damages incurred specifically by coastal fishing communities.

Studying Phailin

Secondary information from government offices, exploratory field visits and discussion with residents of several villages in Gopalpur area (where Cyclone Phailin made landfall in Odisha in 2013) helped in identifying the locations to be considered for the study. Figure 1 shows the track of Cyclone Phailin hitting Gopalpur on the coast of Odisha (IMD 2014). A survey was conducted in six selected villages in Gopalpur area: Gopalpur, Golabandha, Baxipalli, Kotrapur, Thumpsum and Venkatraipur. A total of 300 responses were collected from randomly selected fishing households in order to document their loss and damages. Only the heads of households were interviewed.

They were asked about their losses in income/livelihood and fishing gear, and damage to housing and household well-being such as instances of injury and disease. Further, the time required to start their livelihood, rebuild houses and costs for the same were also noted.

Cyclones Affect Sustainable Livelihood and Income Severely

The survey found that 72.67% of respondents experienced decrease in income after the cyclone (Table 1). This is because a majority of them experienced damages to their boats (69.33%) and fishing nets (65%).

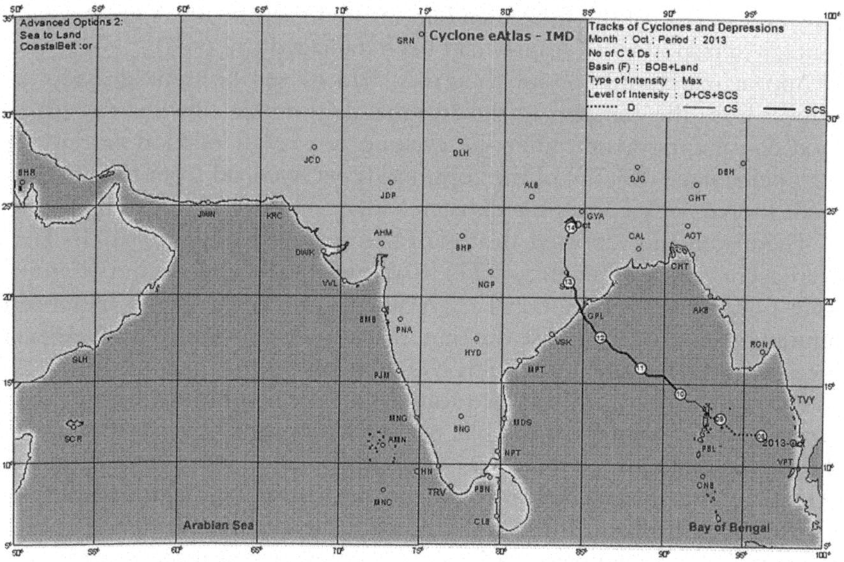

Fig. 1 Track of Cyclone Phailin (IMD 2014)

Table 1 Percentage of households experiencing various damages

		Percentage of households
Livelihood	Decrease in income after cyclone	72.67
	Damage to boat	69.33
	Damage to fishing nets	65.00
Housing	Damage to houses	87.66
Household well-being	Sold household assets to recover	33.33
	Reduced food consumption after Phailin	49.33
	Injury to self	2.33
	Injury to household member (Total responses = 168)	1.79
	Suffered disease by self	0.67
	Suffered disease by household member (Total responses = 168)	0.00
	Household member gone missing (Total responses = 168)	0.60
	Death of household member (Total responses = 168)	0.00

Note: Several households did not provide responses to all the questions, in which case the total number of responses was reduced

However, 73.2% of the fishermen had insured their boats. A greater percentage of respondents' houses had been damaged (87.66%).

Approximately 33.33% of fishermen had to sell household assets to recover from the damages of the cyclone. Around 49.33% had to reduce food consumption right after the cyclone as a result of food scarcity. A high percentage (83.7%) of the community received aid from the government for recovering from the disaster.

There were no reported deaths in the surveyed households. In fact, there were very few fatalities (~21) as a result of the cyclone in the entire state of Odisha (Senapati 2013). A few of the respondents had been injured or suffered from diseases (e.g. malaria) during the cyclone. Also, a member of one of the households had gone missing.

During the survey, it was also found that the community experienced power cuts and damages to public infrastructure such as roads and jetties during the cyclone.

Table 2 lists the time required and costs incurred by the households to recover their livelihood and housing after the cyclone. It shows that a majority of the respondents (57.33%) could go back to fishing within one to three weeks. Approximately 25.67% of the respondents started to fish within a month of the cyclone. Several of them (1.67%) took 7–12 months to start fishing again. The variability in the time taken to go back to fishing may be because of the differences in the damages to gear incurred by the fishermen.

The survey shows that the percentages of fishermen who could return to their previous levels of income within the same time frame ('time taken to go back to fishing') are relatively lower. Only approximately 2% could recover to the same levels of income within a week, compared to 5% who could start fishing in less than seven days. Around 15.84% could recover their income levels in one to three weeks, compared to 57.33% who could return to fishing within this time frame. While all the surveyed fishermen could go back to fishing within a year, it took two years for their income to recover. Thus, although fishermen were able to start fishing relatively sooner after the cyclone, the time taken to recover their income was greater.

The delay in recovering income compared to the time taken to go back to fishing may be because of the time and cost taken to completely recover their fishing gear. A high percentage of the respondents (43.36%) took two to three months to repair or buy new fishing gear. Around 15.93% took a month for the same. A very low percentage (1.77%) could buy or

Table 2 Time taken and costs incurred by the households to recover after the cyclone

	Range	*Percentage of households*
Time taken to go back to fishing	Less than a week	5.00
	1–3 weeks	57.33
	About a month	25.67
	2–3 months	7.33
	4–6 months	2.67
	7–12 months	1.67
Time taken for income to be the same or more than that before the disaster (Total responses = 202)	Less than a week	1.98
	1–3 weeks	15.84
	About a month	11.39
	2–3 months	14.85
	4–6 months	18.81
	7–12 months	35.15
	1–2 years	1.98
Time taken to buy/repair gear (Total responses = 113)	Less than a week	1.77
	1–3 weeks	6.19
	About a month	15.93
	2–3 months	43.36
	4–6 months	13.27
	7–12 months	16.81
	1–2 years	1.77
	2–3 years	0.88
Cost to buy/repair new gear (in INR) (Total responses = 206)	Less than 5000	2.91
	6000–15,000	8.25
	16,000–25,000	11.17
	26,000–35,000	5.83
	36,000–50,000	12.62
	51,000–1lakh	9.22
	>1lakh–2lakh	7.77
	>2lakh–3lakh	5.83
	>3lakh–4lakh	4.85
	>4lakh–5lakh	9.22
	>5lakh–10lakh	9.22
	>10–20lakh	7.77
	>20–30 lakh	2.91
	>30–70lakh	2.43

(*continued*)

Table 2 (continued)

	Range	*Percentage of households*
Time taken to rebuild house (Total responses = 211)	Less than a week	0.00
	1–3 weeks	6.16
	About a month	8.53
	2–3 months	16.59
	4–6 months	16.11
	7–12 months	45.50
	1–2 years	5.69
	2–3 years	1.42
Cost to rebuild house (in INR) (Total responses = 249)	Less than 5000	4.42
	6000–15,000	7.63
	16,000–25,000	5.62
	26,000–35,000	5.62
	36,000–50,000	12.85
	51,000–1lakh	11.24
	>1lakh–2lakh	10.84
	>2lakh–3lakh	19.28
	>3lakh–4lakh	12.45
	>4lakh–5lakh	9.24
	>5lakh–10lakh	0.80

Note: Several households did not provide responses to all the questions, in which case the total number of responses was reduced

repair gear within a week. Further, some respondents also took one to two years (1.77%) and two to three (0.88%) years to replenish their gear.

Most of the respondents had spent INR 36,000–50,000 (12.62%) to repair their gear. A very low percentage (2.91%) incurred gear losses of less than INR 5000. A few respondents also experienced damages of more than INR 20 lakh. The average cost to repair or buy gear was INR 87,450 (1245.20 USD). The total cost of gear repair in the sample was INR 11368500 (161944.28 USD).

Most of the households (45.5%) took around 7–12 months to rebuild their houses. Thus, rebuilding a house was more time-consuming than going back to fishing (57.33% took 1–3 weeks) or repairing gear (43.36% took 2–3 months). Around 12.85% incurred a cost of INR 36,000–50,000 to rebuild their houses. This is similar to the trend followed by the percentage of households incurring costs within the same bracket (INR 36000–50,000) to repair gear. However, a majority of the respondents (19.28%) had to spend around INR 2–3 lakh to rebuild their houses. For several households (0.8%), the costs were greater at around INR 5–10 lakh. Low percentages of the sample experienced costs below INR 35,000. The average cost to rebuild houses was INR 165356.82 (2354.52 USD). Therefore, the average cost of rebuilding houses was far greater (~almost double) than that for repairing/buying gear. The total cost for the same in the sample was INR 37536000 (534700.32 USD). This indicates that the economic value of the damages to houses was higher than that of gear in the community. Thus, institutional support should be provided for the construction of resilient housing in marginalized coastal communities.

Conclusion

This chapter provides an assessment of the loss and damages experienced by the marine fishing community after Cyclone Phailin in 2013. The damages to livelihood, housing and household well-being are reported. Further, time and cost of recovery are also noted. The figures show that large percentages of the community experienced damages. There was a decrease in income of the fishermen and damages to gear after the cyclone. Several fishermen sold household assets to recover from the disaster. Some households experienced impacts on health such as injuries and diseases. Public infrastructure such as electricity and roads was also damaged. Overall, fishermen could start fishing one to three weeks after the cyclone. A small percentage took more than six months to start their livelihood. However, the time taken by the fishermen to recover their income to the same level as before the cyclone was considerably longer. The community also took time ranging from a week to years to replace and repair their damaged gear. Most fishermen took two to three months and spent INR 36,000–50,000 for the same. The time and cost to repair housing was substantially higher compared to gear. Most fishermen repaired their houses over 7 to 12 months and incurred costs of INR 2–3 lakh. Studies on estimation of loss and damages from disasters are growing in the literature and have implications for action and policy on disaster management (van der Geest and Warner 2015). Thus, the present study—by focusing particularly on marine fishing communities—attempts to contribute to this body of limited literature. However, the estimations of this study may not extend to all fishermen and thus future studies could consider providing more robust methodologies and detailed assessments of loss and damages which may be generalized for the entire community.

Acknowledgment The authors are grateful to the Indian Council of Social Science Research (ICSSR) for funding the study.

References

Arunodhaya, Raman, B., Coelho, K., Malar K., Krishnaveni, Dutta, M., Revathi, P., Narayan, P. & Geetha, V. (2016, January). *Sample Survey of Losses Sustained During Chennai Floods*. Available at https://www.cag.org.in/sites/default/files/database/Report%20of%20Survey%20of%20Losses%20Sustained%20during%20Chennai%20Floods_20160113.pdf. Accessed 23 Apr 2019.

Bal, S. (2018). *Community Rebuilds Life After Cyclone Phailin Wrecked Their Fishing Hamlet*. Retrieved October 1, 2018, from https://yourstory.com/2018/01/community-life-cyclone-phailin-odisha/

Crompton, R. P., & McAneney, K. J. (2008). Normalised Australian Insured Losses from Meteorological Hazards: 1967–2006. *Environmental Science and Policy, 11*(5), 371–378. https://doi.org/10.1016/j.envsci.2008.01.005.

Dutta, D., Herath, S., & Musiake, K. (2003). A Mathematical Model for Flood Loss Estimation. *Journal of Hydrology, 277*(1–2), 24–49. https://doi.org/10.1016/S0022-1694(03)00084-2.

Government of India. (2019). *Cyclones & Their Impact in India*. Retrieved May 14, 2019, from https://ncrmp.gov.in/cyclones-their-impact-in-india/

Government of Odisha. (2013). *Cyclone Phailin in Odisha: Rapid Damage and Needs Assessment Report*. Retrieved February 22, 2019, from http://documents.worldbank.org/curated/en/168471468257979992/pdf/838860WP0P14880Box0382116B00PUBLIC0.pdf

IMD. (2014). *Tracks of Cyclones and Depressions*. Retrieved April 30, 2018, from http://www.rmcchennaieatlas.tn.nic.in/Plotting_ByParam.aspx

Kakakhel, K. (2015). *Towards Quantification of Loss and Damage*. Retrieved April 22, 2019, from http://www.lead.org.pk/lead/Publications/32-Towards Quantification of Loss and Damage.pdf

Munich RE. (2019). *Media Information on Natural Catastrophe*. Retrieved May 10, 2019, from https://www.munichre.com/en/media-relations/publications/press-releases/2019/2019-01-08-press-release/index.html

Patankar, A., & Patwardhan, A. (2016). Estimating the Uninsured Losses Due to Extreme Weather Events and Implications for Informal Sector Vulnerability: A Case Study of Mumbai, India. *Natural Hazards, 80*(1), 285–310. https://doi.org/10.1007/s11069-015-1968-3.

Pielke, R. A., Jr., & Landsea, C. W. (1998). Normalized Hurricane Damages in the United States: 1925–95. *Weather and Forecasting, 13*(3), 621–631. https://doi.org/10.1175/1520-0434(1998)013<0621:NHDITU>2.0.CO;2.

Rabbani, G., Rahman, A., & Mainuddin, K. (2013). Salinity-Induced Loss and Damage to Farming Households in Coastal Bangladesh. *International Journal of Global Warming, 5*(4), 400–415. https://doi.org/10.1504/ijgw.2013.057284.

ReliefWeb. (2009). *In-depth Recovery Needs Assessment of Cyclone Aila Affected Areas.* Retrieved May 14, 2019, from https://reliefweb.int/sites/reliefweb.int/files/resources/F6603B7EF22A16B4C125768D004B1190-Full_Report.pdf

Senapati, A. (2013). *UN Felicitates Odisha for Its Disaster Management Model During Phailin. Down to Earth.* Retrieved from https://www.downtoearth.org.in/news/un-felicitates-odisha-for-its-disaster-management-model-during-phailin-43087

The Hindu. (2013). *Phailin: Ganjam Worst Hit, 2.4 Lakh Houses Damaged.* Retrieved October 1, 2018, from https://www.thehindu.com/news/national/other-states/phailin-ganjam-worst-hit-24-lakh-houses-damaged/article5236356.ece

van der Geest, K., & Warner, K. (2015). *What the IPCC 5th Assessment Report Has to Say About Loss and Damage* (UNU-EHS Working Paper No. 21). Bonn.

Wallemacq, P., & House, R. (2018). *Economic Losses, Poverty & Disasters: 1998–2017.* UNDRR and CRED. Retrieved from https://www.unisdr.org/files/61119_credeconomiclosses.pdf

Warner, K., & Van der Geest, K. (2013). Loss and Damage from Climate Change: Local-Level Evidence from Nine Vulnerable Countries. *International Journal of Global Warming, 5*(4), 367–386. https://doi.org/10.1504/ijgw.2013.057289.

Zhang, Q., Wu, L., & Liu, Q. (2009). Tropical Cyclone Damages in China 1983-2006. *Bulletin of the American Meteorological Society, 90*(4), 489–495. https://doi.org/10.1175/2008BAMS2631.1.

Trupti Mishra is an associate professor at the Shailesh J. Mehta School of Management, Indian Institute of Technology Bombay.

Krishna Malakar is a research associate at the Interdisciplinary Program (IDP) in Climate Studies, Indian Institute of Technology Bombay.

Downstream Impact of Melting Glaciers: Climate Change in Nepal and Beyond

Meen B. Poudyal Chhetri

Background

Nepal, a small and land-locked country in South Asia, is exposed to multiple hazards due to the variable geo-climatic conditions, young geology, unplanned settlements, deforestation, environmental degradation and increasing population. Climate change is one of the key factors in the occurrences of various types of disaster. The vast altitudinal variation within a short range of approximately 193 km, from 60 meters to 8848 meters above sea level, makes the country an abundant storehouse of biodiversity and ecological niches with diverse agro-climatic zones ranging from the sub-tropical to the alpine and tundra (Bhattarai and Poudyal Chhetri 2001). However, increasing population, rapid and unplanned urbanization, and other economic activities in vulnerable areas are contributing factors to increased hazards. Hence, Nepal is a global hot spot for several types of disaster.

M. B. Poudyal Chhetri (✉)
Nepal Centre for Disaster Management (NCDM), Kathmandu, Nepal

A. Singh et al. (eds.), *Development in Coastal Zones and Disaster Management*, Disaster Research and Management Series on the Global South, https://doi.org/10.1007/978-981-15-4294-7_20

The Problems

Global warming has become a very serious concern due to its adverse effects on livelihood and the environment. Global climate change is causing rapid melting of snow and glaciers in the Himalayan region. The water from the Himalayan river systems flows into the water basins of nine major rivers serving a total population of almost 1.3 billion. Thus, this mountain range constitutes a major source of water for some of the most populous parts of the world. The region and its water resources play an important role in global atmospheric circulation, biodiversity, irrigated agriculture, potential hydropower, and the production of commodities exported to markets world-wide. Moreover, glaciers are the coolers of the planet and the lifeline of many of the world's major rivers. They contain approximately 75% of the Earth's fresh water and are a source of major rivers. Yet, global warming is melting glaciers not only in the Himalayan region, but also in every region of the world, putting millions of people at risk from floods, droughts and lack of drinking water.

Global Warming and Climate Change

Studies have shown that in the last 100 years the world's average temperature has risen more rapidly than in the last 10,000 years. The scale of temperature rise is increasing. Out of the 10 recorded warmest years in history, nine were recorded during the last decade. The global mean temperature is expected to increase from between 1.4 and 5.8°C over the next 100 years. The adverse effects of such change in global climate are seen in the Himalayas, where glaciers and glacial lakes are posing catastrophic risks. The Himalayan glaciers are retreating at rates ranging from 10 to 60 m per year and many small glaciers (<0.2 sq.km) have vanished. The boundaries of most of the high-altitude valley glaciers in Bhutan, China and Nepal are diminishing quickly. Glaciers in the Himalayan region are thinning faster than elsewhere in the world. If the current situation prevails, the glaciers could disappear by the year 2305. Thus, climate change is shrinking the mountain glaciers and directly affecting the landscape and threatening water supplies all over the world. The Himalayan glaciers can therefore be considered a reliable indicator of climate change, and hence, this is a matter of grave concern.

For the last 10,000 years humans have been living in a remarkably stable climate that has allowed the whole of human development to take place. However, now we see the potential for sudden changes of between 2 and 6 degrees Celsius by the end of the twenty-first century. We do not know what the world will be like at these projected temperatures. We are drifting away from the safe zone (Corell 2007).

Effects of Global Warming in the Nepalese Mountains and Beyond

Mountain regions occupy about a quarter of the global terrestrial land surface and provide goods and services to more than half of their inhabitants. The rise of the Himalayan and Tibetan Plateaus together, which began about 50 million years ago, caused a tremendous impact on the regional and global climate of the world. The Himalayan region has long been recognized as extremely rich in animal and especially plant diversity. The Himalayan watersheds harbor more diverse ecosystems than the Amazon. During the summer season warm moist winds blow inland from the ocean. In the winter, cold dry winds blow from the land toward the ocean. Monsoon circulation involves a change of approximately 180 degrees in the direction of the wind between the summer and winter (Upreti 2008).

Nepal's 83% land mass is mountainous terrain. The wide altitudinal variation along its width gives rise to a steep and rugged topography and extreme relief. Steep and unstable slopes, rugged terrain, active geodynamic processes and intense monsoon rains make the Himalaya an active and fragile mountain range. As the nature of the Himalaya suggests, landslides and debris flows and floods are the main types of water-induced hazards in the region and also in Nepal. These hazards wipe out entire villages, and wash out roads, bridges, canals and hydropower plants, damaging hectares of valuable agricultural land during the monsoon season. Besides substantial economic losses, more than 320 people on average lose their lives in the Nepal Himalaya alone. Other losses from these hazards are on the rise every year. Many factors trigger debris mass movement or debris flows. Among the most common triggers in the Himalaya are prolonged or heavy monsoon rains. Rainfall can lead to mass movement of debris by reducing the internal or binding strength of soil and other materials through three different mechanisms. First, the saturation of soil materials increases the weight of slope materials and creates greater gravitational force. Second, the saturation of soil materials can reduce the cohesive bond of individual soil particles and water can serve as a lubricant along the interface between soil and rock and along the weakness zones of rocks, such as joints, cracks and fault planes. The first two mechanisms often act in combination. The third mechanism is rainfall intensity and duration thresholds that trigger landslides. These have been widely identified in many different climates and geological settings. Caine and Mool (1982) estimate a rainfall threshold of 100 mm day-1 to trigger a landslide leading to a debris flow such as has happened in the Kolphu Khola drainage basin area, central Nepal, in 1980. The intensity and duration of rainfall that can initiate a landslide depends on many factors. However, most landslides and debris flows reported in the Nepal Himalaya are associated with either intense or sustained monsoon precipitation (Dhital 2003, Adhikari and Koshimizu 2005).

There is likely to be increased severity and frequency of monsoonal storms and flooding in the Himalayas, which are expected outcomes of climate change and may significantly alter the area's erosion, river discharge and sediment dynamics. Eventually, this may affect existing hydropower reservoirs, as well as those planned for construction in the Himalayas. Part of the generated sediment may be deposited on agricultural lands or in irrigation canals and streams, which will contribute to deterioration in crop production and in the quality of agricultural lands.

Geo-scientists have found that due to global warming, the number and volume of glacier-lake outburst flood hazards are on the rise. Some of these floods have produced discharge rates of up to 30,000 m3/sec and can run for distances of 200 km (Richardson and Reynolds 2000). Considering the average vertical lapse rate of 6.5 °C per kilometer, it was found that almost 20% of the present glaciated area above 5000 meters altitude is likely to be snow- and glacier-free area with an increase in air temperature of 1 °C. Similarly, a 3 °C and 4 °C rise in temperature could result in ta loss of 58% and 70% of snow and glaciated areas respectively. Such changes are likely to contribute to the faster development of glacier lakes, leading consequently to an increase in potential for glacier-lake outburst flood hazards. Moreover, increase in precipitation by more than 20% is likely to cause significant increase in sediment delivery, and more than 20% increase in annual sediment deposit could be expected in a scenario of 50% increase in annual precipitation (MoPE 2004). The above figures and situations warn us about the effect of global warming and climate change through glacier-lake outburst floods (GLOFs). The occurrence of massive

GLOFs may cause enormous loss not only in Nepal but also in North India, Bhutan and China.

Global climate change is affecting the Himalaya much faster than previously thought, and mountaineers have been the first to notice the changes: more frequent avalanches and more crevasses and exposed rock faces where there used to be snowfields. Cho Oyu, the sixth-highest mountain in the world at 8188 metres (26,864 ft) above sea level, and Chomolungma (Mount Everest) used to be considered the easiest to climb, but have become more difficult in the past 25 years. In the area leading up to the bottom of the Himalaya there are now small lakes and ice slush rivers that form during the spring. The ice walls of the 'Magic Highway' leading to Advanced Base Camp are now half the size of what they used to be 20 years ago. On the Nepal side, Chomolungma (Mount Everest) has also changed dramatically since Edmund Hillary and Tenzing Norgay first climbed it in 1953. 'When my colleague Guy Cotter first guided an expedition to Everest the Hillary Step was completely covered in snow and ice in 2004 it was just rock', says mountaineer Mike Roberts, who has led expeditions to Everest since 2002. 'The entire stretch from the south summit to the true summit is now pure rock'.

Impacts of Climate Change in Nepal

Although Nepal is not contributing to global warming yet climate change has tremendously impacted the glacier ecosystem in the Nepalese Himalayas. Climate change is not just an environmental phenomenon but also an economic, social and political issue in Nepal. From the point of view of climate change, Nepal is among the most vulnerable countries in the world. Himalayan studies are important for four reasons: the role of the Himalaya in global climate change; he impact of global warming on the water balance in high mountain areas; the impact of climate change on flora and fauna; and the socio-economic impact of climate change on subsistence farming.

The major impacts of climate change in Nepal are increased GLOF hazards, increased variability of river runoff, increased sediments, increased evaporation from reservoirs and impacts on watershed. As a result, glacier melt and precipitation patterns are occurring. Nepal has a wide variety of species, and a study has found that 2.4% of biodiversity may be lost due to climate change. It is clear that climate change will affect agriculture, and this is worrying as the majority of the people of Nepal depend on agricultural crops such as rice, maize and wheat. Higher temperatures, increased evapo-transpiration and decreased winter precipitation may result in drought. This should be considered an early warning for food security.

The average maximum temperature in Nepal between 1977 and 1999 increased by 0.9 °C, at a rate of 0.03 °C to 0.12 °C per year, whereas the global average surface temperature rise over the last century was 0.6±0.2 °C (Shrestha et al. 1999), and this is estimated to have risen even higher since then. This is one of the highest registered rates of temperature rise in the world. The observed trend of rising temperature in Nepal is challenging the Intergovernmental Panel on Climate Change projections, as it seems that land areas will warm more rapidly than the global average.

As stated earlier, monsoon climate is predominant in Nepal. Torrential rains during the monsoon render the country highly susceptible to water-induced natural disasters such as floods, landslides, flash floods, debris flows and slope failures. Although rainy days are decreasing, high-intensity rainfall events are increasing, resulting in increase in the magnitude and frequency of water-induced disasters. The potential threat of GLOFs is also growing. GLOFs occur when the moraine damming of a glacial lake suddenly collapses and releases large quantities of water, resulting in a high-velocity surge, causing devastating floods and debris transport downstream.

In the Himalayan region of Nepal glacier lakes are common. Glaciers were formed in the Himalaya between the fifteenth and nineteenth century, during the Little Ice Age (Yamada 1993). A glacier lake originates from a glacier and usually forms at its terminus. According to the International Centre for Integrated Mountain Development (2007), the Nepalese Himalaya has more than 2323 glacier lakes with areas larger than 0.03 sq. km. As a glacier melts, melt water is stored within the lateral and end moraines, creating glacier ice or ice cores and moraines which will continue to grow as the ice melts. The Imja, for example, 'was just a small pond in the 1960s' but it now has a radius of 1 km and stores 2.9 million cubic meters of water (Watanabe et al. 1994). The Koshi River Basin, the Gandaki River Basin, the Karnali River Basin and the Mahakali River Basin contain 1062, 338, 907 and 16 lakes respectively. Dudh Koshi Sub-Basin, the largest basin in Nepal, is also the most densely glaciated region of the country (Bajracharya et al. 2007).

Out of the 2323 glacial lakes, 26 are potentially dangerous. The areas of Upper Barun, Lower Barun, Chamlangtsho, Tsho Rolpa, Sabou, Dudh Kunda, Majang, Imja and Thulagi have been identified as dangerous glacier lakes. These lakes contain huge volumes of water and remain in an unstable condition. As a result, they can burst any time and a natural catastrophe would cause loss of life and property. Approximately 14 such GLOFs have occurred between 1935 and 1991. A GLOF of 1985 caused a 10 to 15 meter-high surge of water and debris to flood down the Bhote Koshi and Dudh Koshi Rivers for 90 kilometers, which swept away a hydropower plant. At its peak, 2000 m3/sec was discharged (Bhattarai and Poudyal Chhetri 2001).

Approximately 20% of the glaciated areas in Nepal above 5000 m are likely to be snow- and glacier-free area with an increase of air temperature of 1 °C. A two-degree Celsius rise in temperature can cause the loss of almost 40% of these areas. Similarly, a 3 °C and 4 °C rise in temperature can result in the loss of approximately 58% and 70% of snow and glacier areas, respectively.

'The rapid melting of Himalayan glaciers will first increase the volume of water in rivers causing widespread flooding', states Jennifer Morgan, Director of WWF's Global Climate Change Program. 'But in a few decades this situation will change and the water level in rivers will decline, meaning massive economic and environmental problems for people in Western China, Nepal, Pakistan and Northern India'.

Conclusions

Due to climate change, glaciers are receding rapidly not only in the Himalayan region. The vast majority of all Himalayan glaciers have been retreating and thinning over the past 30 years, with accelerated losses in the last decade. Glaciers are receding in Africa, the South Pacific, the Arctic, North America, South America, Europe and Antarctica. Glacial retreat is the most visually convincing evidence of climate change, which is not a myth but an unwanted scientific reality. It is high time to realize that adverse impacts of climate change, variability and extremes will be impeding factors to attain the set goals of the world's governments.

Although the United Nations Framework on Climate Change and the Kyoto Protocol allow emission producers to offset their emissions by paying others to carry out emission-reducing activities, it is to be noted that various studies show a pressing need to calculate carbon appropriation, the basis for calculating the impact of ecological imbalance, particularly deforestation. Stopping deforestation and building a healthy environment should be the key issues in climate change policy of every government. This will provide a way for millions of poor people in developing countries to benefit directly. In the same way, such policies will help to reduce deforestation, maintain ecological balance and allow nations to sell credits for successful programs combating carbon dioxide. On the other hand, developed countries that pollute more than the allowed limits under the existing Kyoto accord should buy carbon credits to offset their emission levels and help to fund forest protection plan and programs. We need to work together on reducing CO2 emissions, increasing the use of renewable energy and implementing energy efficiency measures.

Acknowledgments The facts and figures in this chapter are based on the data disseminated from the Ministry of Home Affairs, Kathmandu, Nepal.

References

Adhikari, D. P., & Koshimizu, S. (2005). Debris Flow Disaster in Larcha, Upper Bhotekoshi Valley, Central Nepal. *The Island Arc, 14*(4), 410–423.

Bajracharya, B., Shrestha, A. B., & Rajbhandari, L. (2007). Glacial Lake Outburst Floods in the Sagarmatha Region: Hazard Assessment Using GIS and Hydrodynamic Modeling. *Mountain Research and Development, 27*, 336–344. Kathmandu.

Bhattarai, D., & Poudyal Chhetri, M. B. (2001). *Mitigation and Management of Floods in Nepal.* Nepal: Ministry of Home Affairs, HMG.

Caine, N., & Mool, P. K. (1982). Landslide in the Kolphu Khola Drainage, Middle Mountains, Nepal. *Mountain Research and Development, 2*, 157–167.

Corell, R. (2007). Chairman of the Arctic Climate Change Impact Assessment (ACCIA).

Dhital, M.R. (2003). Causes and Consequences of the 1993 Debris Flows and Landslides in the Kulekhani Watershed, Central Nepal.

ICIMOD. (2007). *Disaster Preparedness for Natural Hazards, Current Status in Nepal.* Kathmandu: International Center for Integrated Mountain Development (ICIMOD).

IPCC. (2007). *Climate Change 2007: The Physical Basis, Contribution of Working Group I to the Fourth Assessment Report of the Intergovernmental Panel on Climate Change.* Cambridge, UK/New York: Cambridge University Press. http://ipcc-wg1.ucar.edu/wg1/wg1-report.html

Ministry of Population and Environment. (2004). *Annual Report*, Kathmandu, Nepal.

Richardson, S. D., & Reynolds, J. M. (2000). An Overview of Glacial Hazards in the Himalayas. *Quaternary International, 65–66*, 31–47.

Shrestha, A. B., Wake, C. P., Mayewski, P. A., & Dibb, J. E. (1999). Maximum Temperature Trends in the Himalaya and Its Vicinity: An Analysis Based on

Temperature Records from Nepal for the Period 1971–94. *Journal of Climate, 12*, 2775–2787.

Upreti, B. N. (2008). *Climate Change and Its Implication in the Himalayan Region*. A Paper Presented at the Queensland University of Technology.

Watanabe, T., Ives, I. D., & Hammond, J. E. (1994). Rapid Growth of a Glacial Lake in Khumbu Himal, Himalaya, Prospects for Catastrophic Flood. *Mountain Research and Development, 14*(4), 329–340.

Yamada, T. (1993). *Glacier Lakes and their Outburst Floods in the Nepal Himalaya* (p. 37). Kathmandu: Published by the Water and Energy Commission Secretariat (WES).

Meen B. Poudyal Chhetri is the President of Nepal Center for Disaster Management (NCDM) and Chairman of The International Emergency Management Society (TIEMS)'s paper Review Committee. He is the Adjunct Professor at the Queensland University of Technology, Brisbane. In the past, he served as the Director of the Department of Disaster Management of the Government of Nepal from 2001 to 2003. Prof. Chhetri also held positions of Speaker, Special Officer, Director, Deputy Regional Administrator, Chief District Officer, Under Secretary and Investigation Officer in various government agencies of Nepal.

Prof. Chhetri has been awarded with **AEI Australia Alumni Excellence Awards 2014** by the Government of Australia on 20 January 2014. He was also awarded with the "Best Paper Award" for his excellent paper titled "**Significance of Cooperation and Coordination in International Disaster Management System**" on 4 October 2013 in Velaux, France.

He has also been awarded with the **"DPNet-Nepal Award 2014"** on 13 October 2014 by the Government of Nepal in recognition to his significant contribution in the field of DRR. He is decorated with various prestigious National Medals like: Janpad Sewa Padak 1991, Mahendra Vidhya Bhusan Padak 1996, Prabal Gorkha Dakshin Bahu 1998, Gaddi Arohan Rajat Padak 2001, Daibi Prakopoddar Padak, 1998, 1999, 2000 and 2001. Prof. Chhetri authored two books namely; "Mitigation and Management of Floods in Nepal" and "Analysis of Nepalese Agriculture." He has also published a number of articles in national and international journals.

PART VI

Preparedness and EWS Technology

Artificial Intelligence Based Early Warning System for Coastal Disasters

Rabindra Lamsal and T. V. Vijay Kumar

Introduction

Abundant sea resources, along with ease of trade and transport, has always been the main reason for human settlements in coastal areas. Although coastal areas appear to be attractive settling grounds, they are also one of the most impacted and disturbed ecosystems world-wide, as they are sensitive to many disasters and risks (Adger et al. 2005). High winds resulting in cyclones and submarine-earthquakes initiating strong tidal waves or tsunami are the major coastal disasters which are responsible for loss of life and infrastructural damage in coastal areas (Finkl and Makowski 2005). Coastal disaster management systems assume utmost importance as means for improving the resilience of the affected populations. It is therefore necessary to have early warning systems (EWSs) in order to address

R. Lamsal (✉) • T. V. Vijay Kumar
School of Computer and Systems Science, Jawaharlal Nehru University, New Delhi, India

Special Centre for Disaster Research, Jawaharlal Nehru University, New Delhi, India

A. Singh et al. (eds.), *Development in Coastal Zones and Disaster Management*, Disaster Research and Management Series on the Global South, https://doi.org/10.1007/978-981-15-4294-7_21

potential disastrous events like in-city flooding due to underwater seismic activity.

Natural hazards are inevitable and addressing them depends on optimum utilization of resources and human settlement in the coastal areas (Diamond and Ordunio 1999; Sidle et al. 2004). More than 0.6 billion people, that is, around 10% of the world's population in 2017, lived in coastal areas, which are less than 10 meters above sea level (United Nations 2017). According to UN Population Division (2001), the population of people living in, and near, coastal areas is likely to reach more than 3 billion by 2025. These areas are more vulnerable to rising sea levels, resulting in inundations, and encroachment of tidal waters into estuaries and river systems. Moreover, coastal areas are also equally exposed to transmission of marine-related infectious diseases. Due to the ongoing trend in human settlements in coastal areas and pressure on the ecosystem, community resilience should form the main focus for sustainable livelihood in those areas (Duxbury and Dickinson 2007). Two-thirds of hazards recorded each year in coastal areas are associated with extreme weather events, such as storms and floods, which are likely to result in more persuasive threats because of anthropogenic changes in the Earth's climate and the rise in sea levels (Adger et al. 2005).

Out of the 1.28 billion population of India, the total population of people living in the coastal districts is 171 million (Database on Coastal States of India 2017). Since almost 14% of people in India reside in coastal areas, there should be proper implementation of possible management measures for addressing the four phases of a disaster, namely mitigation, preparedness, response and recovery.

Disaster mitigation deals with reducing the risk of occurrence of a disaster and its possible consequences. Extreme weather events, such as hurricanes and storms, and natural hazards like underwater earthquakes resulting in tsunamis, become disasters when they directly impact a concentration of human population and infrastructure. This requires efficient and effective governance that can help to mitigate the effects of such hazards through various procedures, which include mapping of vulnerable areas, early warning of potential hazards, enforcement of land use and zoning practices (Godschalk et al. 2000), and insurance programs (Ready.gov 2019).

Disaster preparedness refers to measures taken to prepare for and reduce the overall impact of disasters. According to the United Nations (Kent 1994), disaster preparedness is defined as 'forecasting and taking precautionary measures prior to an imminent threat when advance warnings are possible'. Precautionary measures include research, planning, education and rehearsing.

Disaster response involves the rescuing from immediate danger and the stabilization of the physical and emotional condition of survivors (GOV. UK 2013). This phase of a disaster consists of various activities that include the dissemination of warnings, performing evacuations, search, rescue and damage assessment, providing immediate assistance, and fulfilling basic needs of the affected people as stopgaps until permanent solutions are found.

Disaster recovery relates to a set of policies, tools, and procedures for the assessment, restoration, rehabilitation and reconstruction of disaster systems. The major goal of this phase is to bring back the affected area to normalcy as quickly as possible (UNDP 2010). During disasters such as floods and earthquakes, the recovery process is more hazardous, because it involves health issues of the affected population as a major consideration. Factors to consider during the recovery phase include health tips, general precautions, clothing and personal protective equipment, and electrical/fire hazards (OSHA.gov 2019).

Table 1 lists some of the most powerful seismic events that resulted in the occurrence of tsunamis that impacted the nearby coastal areas and claimed thousands of lives, leaving millions of dollars' worth of infrastructure damaged. The most recently reported highly destructive tsunami was the 2004 Indian Ocean Earthquake and Tsunami, followed by the 2011 Tohoku Earthquake and Tsunami. The 2004 Indian Ocean Earthquake and Tsunami claimed the lives of more than 2, 27,000 people (Telford and

Table 1 Most powerful tsunamis sorted based on event magnitude and casualty

Event	*Magnitude*	*Reported casualties*
2004 Indian ocean earthquake	9.1–9.3 M_w	227,898 dead
2011 Great East Japan earthquake	9.0–9.1 M_w	15,896 dead, 6157 injured and 2537 missing
1755 Lisbon earthquake	8.5–9.0 M_w	60,000 dead
1883 Eruption of Krakatoa	VEI 6	40,000 dead, out of which 2000 deaths due to eruption
1498 Nankai earthquake	8.6 M_S	31,000 dead
1868 Arica earthquake (Peru-Chile coast)	8.5–9 M_w	25,000 dead

Phillips (2011) and NGDC, Natural Hazards Dat (2018)

Cosgrave 2007) in 14 countries, and affected multiple districts of the Indian states of Tamil Nadu, Kerala and the union territory of Puducherry. The 2011 Tohoku Earthquake and Tsunami, which was the most powerful earthquake ever to hit Japan, killed around 18,000 people while leaving 228,863 people homeless.

Such disastrous events put thousands of lives at risk, and thus require the designing of an effective disaster management plan for the people living in these potentially vulnerable areas. For this, EWSs need to be designed to broadcast the relevant information effectively and efficiently, as alarms or warnings, to neighborhoods at risk, during or before such disasters, so that adequate steps can be taken to minimize the loss and damages associated with such disasters. In this chapter, an EWS is proposed that learns from the past underwater seismic data and predicts the extent of horizontal inundation. The proposed EWS pre-processes the relevant features related to the considered data and, using machine learning methods, designs a classifier that can be used to predict, for future data, the extent of horizontal inundation. The proposed EWS is applied to the tsunami dataset and machine learning methods such as Naive Bayes, logistic regression, random forest and artificial neural networks (ANNs) have been used to design the classifier. The performance of the models is compared based on performance measures such as accuracy, precision, recall and F-Measure.

The chapter is organized as follows: EWSs and the related works are discussed in the section 'Early Warning Systems'. The section 'The Proposed EWS' discusses the proposed EWS, followed by experimental results in the section 'Experiments on Tsunami Dataset', and lastly the section 'Conclusion'.

Early Warning Systems

The primary challenge for the researchers in designing EWSs for tsunamis is obtaining information on the precise location of the earthquake and the focal depth, as and when the event occurs. Developing such systems would require instant detection of the magnitude and location of the earthquakes with the installation of near real-time communicating sensor networks consisting of Ocean-bottom Seismometers, real-time Global Positioning System (GPS) devices, and a Tide Gauge system for measuring the change in level of tide waves relative to a reference datum. GPS is the fundamental unit of ESWs because it provides the precise location, that is, latitude and longitude of a region, where events like earthquakes occur. Various studies (Blewitt et al. 2009; Sobolev et al. 2007; Blewitt et al. 2006) discuss in detail the key design specifications of such GPS-based systems, which facilitate detection of earthquakes in near-real-time, for broadcasting warnings to the vulnerable communities.

Some interesting works have been carried out in the design of EWS for earthquakes and tsunamis. An EWS has been designed by the German Research Centre for Geosciences (Rudloff et al. 2009) for determining the location, the magnitude and the depth of an earthquake, which was eventually able to create an alert for a real tsunami event in less than 5 minutes of its occurrence. The system predicts the possibility of the occurrence of tsunami, based on the near-real-time GPS data and the data received from the off-shore systems within a possible minimum time (Falck et al. 2010). Similarly, a tsunami EWS (Liu et al. 2009) has been designed that involves the deployment of sensors in the deep ocean, division of a region of interest into multiple segments and data assimilation from all segments. A decision for a warning or evacuation is made based on the weighted data obtained from the sensors deployed at various segments.

Besides the use of sensor networks for an EWS, social network analysis has also been performed (Chatfield and Brajawidagda 2012) to identify if social platforms like Twitter can act as early warning networks and disseminate information to the vulnerable communities about possible tsunamis. The governmental agency BKMG (Badan Meteorologi, Klimatologi, dan Geofisika), which is responsible for disaster warnings, tweeted about a possible tsunami just 6 minutes and 7 seconds after the occurrence of a powerful earthquake (8.2-moment magnitude scale) on its official Twitter page and, interestingly, it was seen that due to large numbers of people retweeting the same, the tweet reached around 4,102,730 users within 15 minutes of the occurrence of the earthquake.

Currently, only regional tsunami EWSs are in place (Grasso and Singh 2011). The National Oceanic and Atmospheric Organization (NOAA) monitors the seismic stations operated by the Pacific Tsunami Warning System (PTWS), the US Geological Survey and the Alaska Tsunami Warning Center to detect potential earthquakes that might result in tsunamis. The detection of such earthquakes is done depending on the predictors, including the magnitude, focal depth and location of the earthquake. Once a possible tsunami-earthquake is predicted, PTWS issues warnings to the areas that are more likely to be affected.

An EWS that makes use of numerous predictors to provide on-time predictions about possible runup resulting from an underwater seismic activity is proposed and discussed in the next section.

The Proposed EWS

The proposed EWS uses predictors, such as magnitude, depth and location, of an underwater earthquake to predict the extent of horizontal in-city flooding. The overview of the proposed EWS is illustrated in Fig. 1.

The figure shows the source of underwater seismic activity and the vulnerable area or region that is most likely to be affected. The seismic data is generated by the sensor networks deployed in the deep oceans and the tide gauges present on the shores. The sensor network consists of individual units composed of Ocean-bottom Seismometers and near-real-time GPS systems in order to continuously sense for any underwater seismic activity. At the time of an underwater earthquake, the sensor network data at the location of the earthquake, along with the tide gauge historical data of a particular city, is given as input to the machine learning model used by the EWS. The EWS, thereafter, predicts the maximum horizontal in-city flooding for that respective city.

In Rudloff et al. (2009), the features listed in Table 2 related to seismic data, GPS data and tide gauge data are considered. These features, hereafter in this chapter referred to as *tsunamic* data, have been considered in the proposed EWS.

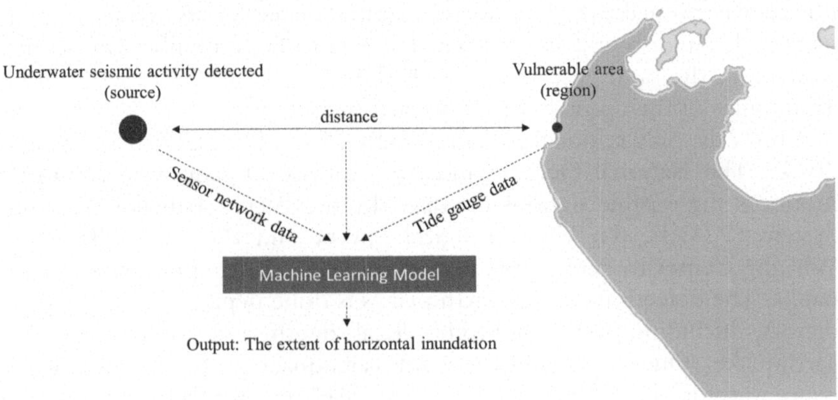

Fig. 1 Illustration of the proposed EWS

Table 2 Features considered in the proposed EWS

	Features
Seismic data	Magnitude, focal depth
GPS data	Latitude, longitude, region source, region runup, distance from source
Tide gauge data	Water height

The proposed EWS takes the values of the features listed in Table 2, related to tsunamic data as input, and produces a classifier that classifies the given tsunamic data instance into *MinorFloodCoverage*, *ModerateFloodCoverage* and *MajorFloodCoverage*, denoting the extent of horizontal inundation as output. The EWS comprises the following steps:

Step 1: Pre-processing tsunamic data

- Remove the instances with missing value(s)
- Normalize the value of features in a common scale, using Min-max Normalization

Step 2: Designing classifier

- Using the normalized values of the features, for the instances selected in step 1, train a classification model that comprises three classes, that is, *MinorFloodCoverage*, *ModerateFloodCoverage* and *MajorFloodCoverage*, and use 10-fold cross-validation thereafter.

The classifier is used to predict the extent of horizontal inundation for a given new instance of tsunamic data. The above steps are discussed in detail in the following sub-sections.

Pre-processing Seismic Data

Data pre-processing is concerned with transforming the raw dataset into a form that can be used for analysis. The representation and quality of the data plays a key role in enhancing the performance of an analytical technique. Due to the presence of unreliable and noisy instances of data, the extraction of knowledge during the training phase becomes difficult (Kotsiantis et al. 2007). Data collected from various sources contain a lot of noise, which makes data pre-processing a complex and time-consuming phase in the knowledge discovery process (Teng 1999). There are four sub-stages that go into data pre-processing, that is, instance selection and outlier detection, dealing with missing values, discretization and data normalization (Kotsiantis et al. 2007). Out of this, missing values and data normalization are dealt with in the proposed EWS. These are discussed below.

Dealing with Missing Values
There are several ways of dealing with missing values in datasets. Instances with missing values can be removed from the dataset, so long as removing those instances does not lead to degradation of the quality of information. Another way is to update the missing values with the most common value of the feature or most common value, specific to its class, of the feature. In addition, missing values can be updated by the average or mean value for a feature. Alternately, regression or classification models can be designed to compute the missing values. In the proposed EWS, the tsunamic data instances with missing values are removed.

Data Normalization
Normalization is concerned with scaling of the value of numeric features to a common scale within a given range, while retaining the same proportion of the difference, among the values, within the range. Data normalization helps in achieving consistency, in terms of computing values within the same range, for all the features. Of the various techniques used for normalization, that is, *Min-Max, Z-Score and Decimal Scaling* (Larose and Larose 2014), the *Min-Max* method has been used for normalizing the tsunamic data. The normalized value, for a given feature, for a given instance using the *Min-Max* method is computed as given below:

$$v' = \frac{v - \min_A}{\max_A - \min_A} \tag{1}$$

where v' is the normalized value and v is the old value, for a feature, of the given instance. min_A and max_A are the minimum and maximum values respectively for the features across all instances.

Designing Classifier

Classification, a supervised machine learning technique, is concerned with identifying a class to which a new observation belongs. A classifier is designed using the existing datasets for which the class labels are already known. Major portions of the existing dataset are used for training the classifier and the remaining portion is used to validate the classifier. Some of the widely used learning algorithms for classification include logistic regression, Naive Bayes, ANNs, support vector machines, k-nearest neighbor and random decision forests (Kotsiantis 2007). The proposed EWS takes the normalized values of the features of the tsunamic data instance and uses that to train a classification model that comprises three classes, that is, *MinorFloodCoverage*, *ModerateFloodCoverage* and *MajorFloodCoverage*, followed by performing a 10-fold cross-validation.

The classifier would aim to predict the extent of horizontal inundation for a given new tsunamic instance by classifying it into one of the above-mentioned classes.

Experiments on Tsunami Dataset

The National Centers for Environmental Information (NCEI) provide two related datasets (National Geophysical Data Centre (NGDC), Global Historical Tsunami Database 2015) that contain information about the source of tsunamis and the locations where the effects of each tsunami were observed in the Atlantic, Indian and Pacific Oceans. The first dataset concerns tsunami source events and contains 49 relevant descriptors including the location, date and time, magnitude, region code, maximum water height, and casualties. The second dataset contains more than 26,000 instances of the runups which were induced by events (recorded in the first dataset) such as an earthquake, a volcanic eruption or a landslide, and contains 37 relevant descriptors including the location of the event, the distance between the source event and the affected area (region), and horizontal inundation.

The events in the source event dataset are classified into various categories, that is, definite tsunami, probable tsunami, questionable tsunami and very doubtful tsunami. Only instances of definite tsunamis have been considered for experimentation. Each instance in the runup dataset is associated with a source event in the source event dataset, with the descriptor event ID acting as the foreign key in the runup dataset. The two datasets were combined with a JOIN operation. The features given in Table 2 are considered in the definite tsunami instances for designing the classifier.

Following step 1, the instances in the tsunami dataset with missing values for any of the features are filtered out. Further, since the descriptors have numerical values in different ranges, the values of all the features are normalized using the *Min-Max* method, discussed in the section 'The

Proposed EWS', in order to scale the values of all the features within the same range, that is, between 0 and 1.

As per step 2, a classifier has to be designed. Predicting a numerical value as the maximum in-city flooding distance is a regression problem, although there is a lower possibility of predicting the flooding distance close to the real value because of an unexplored dimension, that is, the slope of the shoreland. The NCEI datasets do not provide the slope or inclination of the shoreland for the historical runup events. Even if this problem is considered a regression problem, this unexplored dependent variable is likely to induce large variation between predicted and real values. A better solution is to convert the regression problem to a classification problem with classes, as illustrated in Table 3. The effect of the unexplored dimension on the dependent variable can be incorporated by predicting the class for the extent of in-city flooding distance.

Since there is no classifier which works best for all types of classification problems, the pre-processed tsunami dataset was given as input to numerous classification-based learning algorithms, that is, Naive Bayes, logistic regression, random forests and ANNs.

The Naive Bayes classifier is a conditional probabilistic learning model (Langley et al. 1992). Given an instance of a problem $x = (x1, x2, x3, \ldots)$ with n independent variables for classification, the model assigns probabilities to the instance for each class C_i.

$$p(Ci|x1, x2, x3, \ldots, xn) \tag{2}$$

The above conditional probability model is computed based on Bayes's Theorem, as given below:

$$p(Ci|x) = \frac{p(Ci) * p(x|Ci)}{p(x)} \tag{3}$$

Table 3 Categorization of in-city flooding

Classes	*In-city flooding (in meters)*
MinorFloodCoverage	Below 2000
ModerateFloodCoverage	2000–4000
MajorFloodCoverage	4000 and beyond

where $p(C_i|x)$ is the probability that instance x belongs to class C_i, $p(x|C_i)$, the probability of an instance x given class C_i, $p(C_i)$ is the probability that any instance belongs to class C_i and $p(x)$ is the probability of an instance x irrespective of the object it belongs to.

Logistic regression (Cox 1958; Walker and Duncan 1967) is one of the most widely used machine learning algorithms for classification tasks. Given a set of independent variables (x_1, x_2, x_3, ..., x_n), a dependent variable (y) and the parameters (slopes) b_i of the model, the logistic model takes a form which is similar to linear regression:

$$y = b_0 + b_1 x_1 + b_2 x_2 + \ldots + b_n x_n \tag{4}$$

Now, the Logistic Function, that is, the sigmoid function, takes the real value of y and output values strictly between 0 and 1, as probability, using the equation given below:

$$p(y) = \frac{1}{1 + e^{-(y)}} \tag{5}$$

where $p(y)$ is the probability of the dependent variable.

Random decision forest (Ho 1995), which is an ensemble learning method used for both regression and classification problems, constructs multiple decision trees using random subsets of the training dataset and assigns the new tsunamic instance to the class that gets the majority votes. It constructs a given number of trees, say n, by randomly selecting instances from the dataset. These n trees, thereafter, are used to classify a new instance. For this, distribution of votes for all outcome classes is computed and the instance is classified to the class having a majority of the votes.

Artificial Neural Network is a machine learning approach (Haykin 1994) which is inspired by biological neurons. ANN is a network of artificial neurons, where each neuron receives inputs, computes the activation function based on the inputs, to change its state, and finally computes the outputs, based on the inputs and their activation. First, neural network structure is defined by initializing the number of nodes in the input and the hidden layers. This is followed by initializing the biases and weights. Next, the forward propagation, the backward propagation and parameter updation using the gradient descent method are carried out until the loss value reaches a saturation level.

The above techniques have been applied to the normalized tsunami dataset. The widely used performance measures applied to assess the performance of the classifiers are accuracy, precision, recall and F-Measure (Sokolova and Lapalme 2009). These are briefly discussed below.

Accuracy is the ratio of correctly classified instances to total number of instances, as given below:

$$Accuracy = \frac{TP + TN}{TP + TN + FP + FN} \tag{6}$$

where *TP* is true positive, *TN* is true negative, *FP* is false positive and *FN* is false negative. *Precision* is the ratio of correctly predicted positive instances to total predicted positive instances, as given below:

$$Precision = \frac{TP}{TP + FP} \tag{7}$$

Recall is the ratio of correctly predicted positive instances to total number of instances in the class, as given below:

$$Recall = \frac{TP}{TP + FN} \tag{8}$$

F-Measure is the weighted average of precision and recall, as given below:

$$F - Measure = \frac{2 * Precision * Recall}{Precision + Recall} \tag{9}$$

The results of experiments are given in Table 4. It can be inferred from the table that the ANN outperformed all other classifiers, in terms of accuracy, correctly classifying 83.33% of the instances of the tsunamic dataset. Moreover, for ANN, the weighted average of precision and recall, that is, F-Measure, for class *MinorFloodCoverage* and class *ModerateFloodCoverage*, was 0.82 and 0.88 respectively. The precision and recall for class *MajorFloodCoverage* were relatively lower with regard to other classes. The model predicts the category of in-city flooding distance, based on the strict range of each category. There were some instances in the dataset, for example, 4091 meters, 4257 meters and 4267 meters, for which the

Table 4 Performance of various classifiers

Classifier	*Accuracy*	*Precision*	*Recall*	*F-Measure*	*Class*
Naive Bayes	76.38%	0.741	0.541	0.625	*MinorFloodCoverage*
		0.847	0.922	0.883	*ModerateFloodCoverage*
		0.368	0.412	0.389	*MajorFloodCoverage*
Logistic regression	81.94%	0.824	0.757	0.789	*MinorFloodCoverage*
		0.830	0.922	0.874	*ModerateFloodCoverage*
		0.700	0.412	0.519	*MajorFloodCoverage*
Random forests	82.63%	0.771	0.730	0.750	*MinorFloodCoverage*
		0.853	0.900	0.876	*ModerateFloodCoverage*
		0.786	0.647	0.710	*MajorFloodCoverage*
Artificial neural network	83.33%	0.879	0.784	0.829	*MinorFloodCoverage*
		0.847	0.922	0.883	*ModerateFloodCoverage*
		0.615	0.471	0.533	*MajorFloodCoverage*

model incorrectly classified to class *ModerateFloodCoverage*, which represents the range 2000–4000. Instead, they should have been classified to the class *MajorFloodCoverage*. Therefore, the prediction model has a moderate possibility of incorrectly classifying the instances that fall near the left edge of the range of the class *MajorFloodCoverage* to class *ModerateFloodCoverage*. Further, random forest performed the next best, followed by logistic regression. Naive Bayes performed comparatively poorer among the four classification methods.

Conclusion

Coastal areas are vulnerable to numerous natural disasters like high winds resulting in cyclones and underwater earthquakes initiating strong tidal waves or tsunamis. The focus of this chapter was to develop an EWS which is capable of predicting the maximum horizontal in-city flooding distance and accordingly raising an alarm to the areas that are most likely to be affected by the underwater seismic activity. In this chapter, a classifier was trained using four widely used classification-based learning algorithms, that is, Naive Bayes, logistic regression, random forests and ANNs. Among these, the ANNs outperformed all other classification algorithms by correctly classifying comparatively higher percentage instances of the tsunami dataset.

As a whole, the proposed EWS is capable of predicting the extent of horizontal inundation, for a given new tsunamic data instance, by classifying it into classes based on the extent of horizontal inundation, that is, classes *MinorFloodCoverage*, *ModerateFloodCoverage* and *MajorFloodCoverage*. This would enable dissemination of the extent of likely flooding information to communities at risk and the various governmental agencies, so that timely and adequate measures could be taken to reduce the potential harm or damage that can be caused by floods.

References

Adger, W. N., Hughes, T. P., Folke, C., Carpenter, S. R., & Rockström, J. (2005). Social-Ecological Resilience to Coastal Disasters. *Science, 309*(5737), 1036–1039.

Blewitt, G., Kreemer, C., Hammond, W. C., Plag, H.-P., Stein, S., & Okal, E. (2006). Rapid Determination of Earthquake Magnitude Using GPS for Tsunami Warning Systems. *Geophysical Research Letters, 33*(11).

Blewitt, G., Hammond, W. C., Kreemer, C., Plag, H.-P., Stein, S., & Okal, E. (2009). GPS for Real-Time Earthquake Source Determination and Tsunami. *Journal of Geodesy, 83*, 335–343.

Chatfield, A. T., & Brajawidagda, U. (2012). Twitter Tsunami Early Warning Network: A Social Network Analysis of Twitter Information Flows. In *23rd Australasian Conference on Information Systems* (pp. 1–10). Deakin: Deakin University.

Cox, D. R. (1958). The Regression Analysis of Binary Sequences. *Journal of the Royal Statistical Society: Series B: Methodological, 20*(2), 215–242.

Database on Coastal States of India. (2017). Retrieved from Centre for Coastal Zone Management and Coastal Shelter Belt. http://iomenvis.nic.in/index2.aspx?slid=758&sublinkid=119&langid=1&mid=1

Diamond, J. M., & Ordunio, D. (1999). *Guns, Germs, and Steel*. UK: Penguin.

Duxbury, J., & Dickinson, S. (2007). Principles for Sustainable Governance of the Coastal Zone: In the Context of Coastal Disasters. *Ecological Economics, 63*(2–3), 319–330.

Falck, C., Ramatschi, M., Subarya, C., Bartsch, M., Merx, A., Hoeberechts, J., & Schmidt, G. (2010). Near Real-Time GPS Applications for Tsunami Early Warning Systems. *Natural Hazards and Earth System Sciences, 10*, 181–189.

Finkl, C. W., & Makowski, C. (2005). *Encyclopedia of Coastal Science*. Cham: Springer International Publishing.

Godschalk, D. R., Norton, R., Richardson, C., & Salvesen, D. (2000). Avoiding Coastal Hazard Areas: Best State Mitigation Practices. *Environmental Geosciences, 7*(1), 13–22.

GOV.UK. (2013). *Emergency Response and Recovery*. London: Cabinet Office, Civil Contingencies Secretariat. Retrieved from GOV.UK: https://assets.

publishing.service.gov.uk/government/uploads/system/uploads/attachment_data/file/253488/Emergency_Response_and_Recovery_5th_edition_October_2013.pdf

Grasso, V. F., & Singh, A. (2011). Early Warning Systems: State-of-Art Analysis and Future Directions. *Draft report, United Nations Environment Programme, 1*. Nairobi available at https://na.unep.net/siouxfalls/publications/Early_Warning.pdf

Haykin, S. (1994). *Neural Networks* (Vol. 2). New York: Prentice Hall.

Ho, T. K. (1995). Random Decision Forests. In *Proceedings of 3rd International Conference on Document Analysis and Recognition* (Vol. 1, pp. 278–282).

Kent, R. C. (1994). *Disaster Preparedness*. Geneva: UNDP/DHA Disaster Management Training Programme.

Kotsiantis, S. B. (2007). Supervised Machine Learning: A Review of Classification Techniques. In *Proceedings of the 2007 Conference on Emerging Artificial Intelligence Applications in Computer Engineering: Real Word AI Systems with Applications in eHealth, HCI, Information Retrieval and Pervasive Technologies* (pp. 3–24). Amsterdam: IOS Press.

Kotsiantis, S. B., Kanellopoulos, D., & Pintelas, P. E. (2007). Data Preprocessing for Supervised Learning. *International Journal of Computer, Electrical, Automation, Control and Information Engineering, 1*(12), 4091–4096.

Langley, P., Iba, W., & Thompson, K. (1992). An Analysis of Bayesian Classifiers. In *Proceedings of the Tenth National Conference on Artificial Intelligence (AAAI'92)* (pp. 223–228). San Jose: AAAI Press.

Larose, D. T., & Larose, C. D. (2014). *Discovering Knowledge in Data: An Introduction to Data Mining*. Hoboken: John Wiley & Sons.

Liu, P. L.-F., Wang, X., & Salisbury, A. J. (2009, September 4). Tsunami Hazard and Early Warning System in South China Sea. *Journal of Asian Earth Sciences, 36*(1), 2–12.

NGDC. (2015). *Global Historical Tsunami Database*. https://doi.org/10.7289/V5PN93H7

NGDC. (2018). *Natural Hazards Data*. Retrieved from National Centers for Environmental Information | National Oceanic and Atmospheric Administration. https://www.ngdc.noaa.gov/hazard/

OSHA.gov. (2019). *Fact Sheet on Natural Disaster Recovery*. Retrieved from Occupational Safety & Health Administration, U.S. Department of Labor. https://www.osha.gov/OshDoc/cleanupHazard.html

Phillips, C. (2011, March 16). *The 10 most Destructive Tsunamis in History*. Retrieved from Australian Geographic: https://www.australiangeographic.com.au/topics/science-environment/2011/03/the-10-most-destructive-tsunamis-in-history/

Ready.gov. (2019). *Risk Mitigation*. Retrieved from https://www.ready.gov/risk-mitigation

Rudloff, A., Lauterjung, J., Münch, U., & Tinti, S. (2009). Preface "The GITEWS Project (German-Indonesian Tsunami Early Warning System)". *Natural Hazards and Earth System Sciences, 9*, 1381–1382.

Sidle, R. C., Taylor, D., Lu, X., Adger, W., Lowe, D., De Lange, W., Newnham, R., & Dodson, J. (2004). Interactions of Natural Hazards and Society in Austral-Asia: Evidence in Past and Recent Records. *Quaternary International, 118*, 181–203.

Sobolev, S. V., Babeyko, A. Y., Wang, R., Hoechner, A., Galas, R., Rothacher, M., Sein, D. V., Schröter, J., Lauterjung, J., & Subarya, C. (2007). Tsunami Early Warning Using GPS-Shield Arrays. *Journal of Geophysical Research, 112*, B08415. https://doi.org/10.1029/2006JB004640. Available from: https://www.researchgate.net/publication/248803935_Tsunami_early_warning_using_GPS-Shield_arrays. Accessed 24 Aug 2020.

Sokolova, M., & Lapalme, G. (2009). A Systematic Analysis of Performance Measures for Classification Tasks. *Information Processing & Management, 45*(4), 427–437.

Telford, J., & Cosgrave, J. (2007). The International Humanitarian System and the 2004 Indian Ocean Earthquake and Tsunamis. *Disasters, 31*(1), 1–28.

Teng, C.-M. (1999). Correcting Noisy Data. In *Proceedings of the Sixteenth International Conference on Machine Learning* (pp. 239–248). San Francisco: Morgan Kaufmann Publishers Inc.

Tsunami early warning using GPS-Shield arrays. Available from: https://www.researchgate.net/publication/248803935_Tsunami_early_warning_using_GPS-Shield_arrays [accessed Aug 24 2020].

U. N. Population Division. (2001). *World Population Prospects: The 2000 Revision, Volume III*. New York: United Nations.

UNDP. (2010). *Disaster Risk Reduction and Recovery*. New York: Bureau for Crisis Prevention and Recovery.

United Nations. (2017). *Factsheet: People and Oceans*. New York: The Ocean Conference. Retrieved from https://www.un.org/sustainabledevelopment/wp-content/uploads/2017/05/Ocean-fact-sheet-package.pdf.

Walker, S. H., & Duncan, D. B. (1967). Estimation of the Probability of an Event as a Function of Several Independent Variables. *Biometrika, 54*(1/2), 167–179.

Rabindra Lamsal is a technology researcher at the Special Centre for Disaster Research, and an MTech from the School of Computer & Systems Sciences, JNU, New Delhi.

T.V. Vijay Kumar is a professor and dean of School of Computer & Systems Sciences, and a concurrent professor at the Special Centre for Disaster Research, JNU, New Delhi.

PART VII

Path Ahead

Critical Coastal Planning to Prevent Coastal Elegy

Amita Singh

Latest Projections on Increasing Coastal Vulnerability to Disasters

The latest projections on sea level rise and the percentage of population affected alerts governments and communities to ensuring critical coastal preparedness along the coastal rim. Climate change, despite being just one of the reasons for sea level rise, is disproportionately highlighted in the disaster literature since it diverts funds and attention from identifying human frailties, lapses and administrative accountability. No single country or institution is either a cause of climate change or a preventive. Climate change is globally caused and can be prevented through concerted global action. This is the reason why disaster management has an Act and a committed administrative structure to mitigate and prevent disasters. Since the time of NASA scientist James Hanson's testimony to the US Senate in 1988, on threats related to climate change yet by getting embroiled into

A. Singh (✉)
Centre for the Study of Law & Governance, Jawaharlal Nehru University, New Delhi, India

A. Singh et al. (eds.), *Development in Coastal Zones and Disaster Management*, Disaster Research and Management Series on the Global South, https://doi.org/10.1007/978-981-15-4294-7_22

it would divert attention from a much required immediate institutional preparedness which should be undertaken by the government. Climate change experts provide projections which can and should be incorporated into planning and preparedness exercises. The science of hazard assessments and warnings about impending disasters have become clearer and more insightful since Hanson disclosed the impacts of climate change, and today planning for mitigation and prevention is possible.

A formidable recent revelation on coastal data should force SDMA officials to take responsibility. The new estimates are three times higher on population affected by sea level rise, and when combined with 1.5 times faster coastal erosion more than 36 million people would be affected by annual flooding by 2050 as against the previously estimated 5 million only. According to an IIT Bombay joint study[1] with the National Centre for Earth Science Studies, Thiruvananthapuram, coastal erosion is currently 1.5 times faster than during the last three decades, and percentage of land loss in each coastal state is as follows.

1. **Land management** in the coastal rim states can increase vulnerability of both the Eastern and the Western Ghats. Coastal land use, if poorly managed, speeds up erosion, and an important issue to consider in preparedness is the urgent need to address land loss or erosion through new coastal conservation technologies and any drastic change warrants wartime preparation. Table 1 shows some drastic changes in the 26 years assessed in the study. West Bengal is the

Table 1 Outcome of study of 6632 kms shoreline distributed in nine states during 1990–2016

Sl. No	*States*	*Land loss (high erosion)*	*Land gain*
1	West Bengal	99.05	16.46
2	TN	17.92	15.37
3	Kerala	7.77	11.13
4	AP	44.83	53.84
5	Gujarat	29.17	61.21
6	Maharashtra	5.49	5.84
7	Karnataka	2.59	3.38
8	Odisha	26.90	63.45
9	Goa	0.55	0.82

Source: Kankara et al. (2018, pp. 15–17)

> most vulnerable and suffers maximum land loss but the study observes that 'apart from Kerala coast, other states on the west coast of India fall in stable condition'(Kankara et al. 2018, p. 15) (Table 1). This statement is significant as even West Bengal and Tamil Nadu join the most vulnerable list with the highest land loss percentage. The study, being a Geographical Information System (GIS)-based monodisciplinary investigation, has not taken into account developmental indicators including population, demography, infrastructure, livelihood and ecosystems which justify urgent attention based on transdisciplinary indicators. A social science-based approach attaches all socio-economic indicators to the scientific findings to make a clearer assessment of levels of vulnerability and to find ways to improve access through participatory governance.

A much worse than estimated scenario is brought out by the latest study by Kulp and Strauss (2019) which reveals triple the previous estimates of affected population in coastal rims of the world. According to this report, most of the coastal areas within India alone would be submerged permanently under water and the number of people affected would rise to 44 million. It further says that Mumbai, Kolkata, Odisha, Gujarat, Surat and Sundarbans in Bengal would be badly affected. The coasts of Kerala and Tamil Nadu are already facing stress and instability to a greater degree than other coastal states.

Kulp and Strauss (2019) have measured coastal elevation with a new innovative methodology. Any projection of people affected by sea level rise depends upon how one measures the elevation. So far this has been measured by NASA's Shuttle Radar Topography Mission, but it showed errors in estimates of coastal elevation by a height of 6 ft. and up to 15 ft. in some places around the world. The authors of the report then used a machine learning-based Digital Elevation Model to rectify the error. The results shocked them as much as the remote sensing results on forest degradation shocked the Indian government in the 1970s. The report makes it very clear that a policy of 'limited mitigation' and 'no adaptation' would sink major cities falling below the tideline, especially the bustling city of Mumbai, all the way up past Vasai and Nala Sopara to the north, and Navi Mumbai and Dombivili to the east. This also applies to areas which are already low lying such as the Diamond Harbour and the Sundarbans in Kolkata to the city's northern suburbs and most of the coastline of Kerala. Employing CoastalDEM methodology shows that the number of people living below projected high tide lines will increase to 190 million people, a median increase of just 80 million, by the year 2150, and this triples to 630 million people by the year 2100 and up to 340 million by the year 2150. According to the report more than 1 billion people now occupy land which is less than 10 meters above current hightide lines. The vulnerable population is only going to increase with the sea level continuing to rise,[2] but even if climate change were not a factor, this population would have been affected due to mismanagement of coastal land areas and disrespect for the laws of nature.

2. **Location of sensitive infrastructure** such as dams and nuclear power plants, which are mostly constructed in the coastal states, need skilled, efficient and strategic attention. Dam management indicates critical disadvantages for coastal states. Some of the largest reservoirs with a height of 100m and above and a stored capacity of 1 billion cubic meters are located in Tamil Nadu, Maharashtra, Kerala, Karnataka, Odisha and Gujarat. Kerala (38,863 km2) and Tamil Nadu (130,060 km2) have much smaller land areas as compared to other coastal states, that is, Maharashtra (307,713 km2), Gujarat (196,024 km2) and Karnataka (191,791 km2), but have a more impressive number of dams and large reservoirs, for example Kerala has around 80 dams over 44 rivers with 43 large dams, while Tamil Nadu has around 30 large and small dams covering the key state rivers Noyyal, Bhavani, Kaveri, Vaigai and Thamirabarani. Maharashtra has 2354 big dams. The fragile coastal ecology of the Western Ghats has been damaged and its resilience has decreased due to massive land grab for construction in fragile ecological regions. The Kerala government passed an ordinance called the Kerala Forest (Vesting and Management of Ecologically Fragile Lands) Amendment Act, 2009 which released a large tract of land for construction and cultivation in defiance of ecological requirements.
3. **The age of the dams in the coastal belt** is also a cause of impending disaster. Kerala's oldest dam, Mullaperiyar, was built in 1855 and its largest dam, Idukki, in 1955. Andhra Pradesh's main barrages, such as Dowleswan, Prakasham and Sunkesula, were all built before 1861 over the Godavari, Krishna and Tungabhadra Rivers respectively. In Maharashtra the situation is similar, and in fact it is worse, as the Dhamapur dam over the river Malvan in the

Sindhudurga area was built in 1530. More than 40 dams in Maharashtra were built between 1860 and 1927.The dams around Mumbai, Pune, Ramtek, Solapur, Nashik and Salara are particularly vulnerable, and these areas need urgent attention for disaster preparedness. Gujarat, with more than 200 dams, offers a guide for the operation of gates and Rule Curve Levels for Irrigation Projects at Narmada Water Resource Department but without much ground-level preparedness. Odisha' largest dam, Hirakud over the river Mahanadi, was constructed in 1947.

4. **Management of nuclear power plants at the coasts is a challenge for disaster governance and institutions.** The largest atomic power plants with more than 2000 MW capacity are spread around Tamil Nadu (Kundakulum and Kalpakkam nuclear reactors), Karnataka (Kaiga nuclear reactor), Gujarat (Kakarapar) and Maharashtra (Tarapore's Bhabha Atomic Power Plant) and call for a governance system to manage critical infrastructure. India has a total of 22 nuclear reactors with an installed capacity of 6780 MW which produce a total of 30,292.91 GWH electricity.[3] A near-emergency governance level appears indispensable now as results of investigations into Japan's infamous March 2011 disasters emerge from the nuclear power plants of Fukushima Daiichi, Onagawa, Tokai and Rokkasho following the Tohoku earthquake and the enormous tsunami that followed. The startling facts which were brought out by the Nuclear Accident Independent Investigation Committee of the National Diet of Japan have alerted the operators and the governments accountable for the safety of these power plants, as well as citizens in the vicinity. These findings exposed the presence of both maladministration and conflict of interest. The operator Tokyo Electricity Power Company was aware of the possibility of an imminent accident at the plant site yet failed to put in place the basic safety requirements of risk assessment, collateral damages and evacuation plans. Consequently, a 20 km area around the plant had to be declared an evacuation zone and 154,000 people were evacuated, but even this could not prevent the radioactive contamination of both the air and the Pacific Ocean into which the radioactive isotopes escaped and will continue to float for centuries to come. The cause of this laxity was later exposed[4] in many findings, including those of the International Atomic Energy Agency, which revealed an inherent conflict of interest in the management of

nuclear power plants. The regulation and promotion of the nuclear power industry was vested in the same government department. This is similar to the conflict of interest exposed in the '2018-Kerala Flood Report'[5](p. 26) on hydro-electricity management and the operation of dams both in the hands of the State Minister M.M. Mani. Governments continue to be unaccountable for their lack of preparedness, negligence regarding warnings and managerial ethics, and this remains the least researched area in the disaster literature, overshadowed by hegemonic science models.

5. **Coastal land,[6] which is ecologically fragile,** is a threatened by political transactions comparable to the battles over common property resources in rural areas inland. There needs to strong political will to protect the fragile ecological areas so endemic to the coastal rim. The Western and Eastern Ghats are covered with dense and rich tropical biodiversity and wildlife. The Western Ghats constitute one of the eight biodiversity hotspots but has been extensively quarried, constructed on and ecologically destroyed to the extent that less than one-third of forested land remains, and this is scattered and fragmented, with very few coastal mangroves, corals and grasses to help in carbon sequestration and supporting marine life. The value of ecosystem services[7] provided by these biomes of tropical forests across the world stands at USD33 trillion per year as compared to global GNP of USD18 trillion only. Much of the land in the Western Ghats is degraded, eroded and overexploited, with incontiguous wildlife corridors which have reduced the sustainable gains otherwise possible through in situ conservation. The information obtained from across the Western Ghats on fragile land management suggests a severe impending disaster. In Kerala, the fragile land management framework is continuously being diluted to suit vested interests of cultivators, farmers and land grabbers. An ordinance was brought in 2009 to amend the Kerala Forest (Vesting and Management of Ecologically Fragile Lands) Act, 2003,and then the Oommen V Oommen Panel was constituted which scrapped the EFL Act at the demand of the vested lobbies in Idukki, Wayanad and Kozhikode. The state government justified the Panel report on the grounds that it wanted to alleviate the fears among these people settled on fragile land after the Centre's Draft Notification issued by the Ministry of Environment, Forest and Climate Change accepted the Kasturirangan Report. The grounds on which the state government accepted the

Oommen V Oommen Panel Report indicates that the government either had no understanding of the relation between fragile land ecology and human security or were working to pacify the political vote bank with eyes wide shut. The Kasturirangan Report affected 123 villages in 12 of the 14 districts by declaring them Ecologically Sensitive Areas (ESA), and the local inhabitants feared that this ESA would soon be converted to EFL and that they would be shifted out.[8] The Labour Unions, Planters Association and Estate Owners came together to highlight the employment and livelihood of 40,000–50,000 people over 640,000 ha land. The 2018 Kerala Floods Report (2018) found the politics of greed and politically protected cultivators to be one of the major factors in repeated flood disasters. Fragile land ecology conservation is one of the major arguments for the protection and conservation of coastal areas from disasters. Some interesting and holistic interpretations of fragile land areas have been offered in various judgments pronounced by the Court. In T.N. Godavarman Thirumulpad vs. Union of India & Ors. ((1997) 2 SCC 267) and in T.N. Godavarman Thirumulpad vs. Union of India & Ors. ((2006) 1 SCC 1) interpreting Section 2 of the Forest Conservation Act 1982, it was held that even those areas which had been denuded, degraded or cleared would still be treated as forests. In compliance with the above law, clearance of forest does not become a reason for construction. Following this interpretation, all such areas which are cleared of forest can only be used for in situ conservation as per the Management Plan under Section 16 of the Act. However, when the Kasturirangan Report emphasized the same principles for protecting the Western Ghats the state government tried amending the law which reflected the need to reclaim fragile land. Sustainable development demands that biodiversity conservation and wildlife protection be practiced and enforced simultaneously. This was also emphasized in State of Gujarat vs. Mirzapur Moti Kureshi Kassab Jamat ((2005) 8 SCC 534), where it was shown that environment, forest and wildlife are interrelated and interdependent; they protect one another.

Modern Technology and Accountability in Disaster Management

Geospatial technology, which combines the three important disciplines of GIS, Remote Sensing and Artificial Intelligence, has produced better records of the causes and consequences of human and machine failures. Many independent organizations are applying these technologies to be prepared for disasters. Recently a Geospatial Conservation Atlas was created by the group Nature Conservancy (geospatial.tnc.org). Similarly, another group called the Urban Design Associates and the *MIT Technology Review* have studied recovery and rebuilding from the massive fires which engulfed the Sierra Nevada foothills in California in 2018–19. Professor Robin Murphy[9] utilizes robotics in multiple spheres of disaster preparedness and rescue operations (ow.ly/10gR9T). It is time for these developing technologies to also be used to set standards for accountability of institutions and administrators, especially the heads of national-, state- and district-level disaster management agencies.

The law has always made accountability fuzzier. The Disaster Management Act 2005 also lacks stringent accountability measures to take punitive measures against those who create disasters, such as legislators for harmful laws, builders for unauthorized construction, administrators for lack of preparedness in the face of disasters, dam managers for floods, state and central governments for diverting or blocking mitigation funds, and lastly insurance companies for failing to respond appropriately during disasters. CRZ laws provide a basic framework for institutional and administrative accountability for coastal damages. CRZ laws ensure protection of coastal areas and conservation in resilience building. CRZs are supported by some of the most important international conventions regarding ecologically fragile or disaster -vulnerable lands:

1. Ramsar Convention on Wetlands
2. Convention on Biological Diversity
3. UNECSO's Man and the Biosphere Programme
4. Convention Concerning the Protection of World Cultural and Natural Heritage

These international programs and conventions to which India is a signatory are part of domestic law. For example, the Supreme Court in T.N. Godavarman Thirumulpad v. Union of India & Ors. ((2002) 10

SCC 606) held that the Convention on Biological Diversity is part of our domestic law. States are duty bound to implement the provisions of these international conventions. National compliance with these laws is needed not only under the Customary International Law but also under the laws made by the judiciary of the country, which in view of Article 141 of the Constitution[10] of India is the law of the land. The legal framework becomes so wide and scattered that accountability is difficult to enforce.

Remote sensing, which mostly consists of aerial sensor technologies, has been a tool of accountability which has changed the whole scenario of decision making across institutions and is involved in planning sustainable futures. Forest and vegetation mapping through remote sensing and GIS technologies enables planners to conduct accurate, timely and cost-effective forest resource evaluations. For example, according to the National Forest Policy, 33% of the total geographical area of a country should be under forest cover, with 66% for hills and 20% for plains. Remote sensing and satellite imagery has exposed and shaken planners from their comfort zones by identifying most areas falling much below the required cover and thereby becoming a cause of landslides, floods and hurricanes. Remote sensor-based accountability systems are being applied in fire detection, firefighting, detection of human and nonhuman bodies, damages, varieties of vulnerability, crime prevention during disasters, and lastly and most important in early warning communication and timely alerts.

Planners and builders generally overstep the carrying capacity of ecosystems, leading to the elimination and destruction of these ecosystems. In conventional systems, governments have maximized data in order to placate their political vote bases and have justified denotification of forest areas, groundwater withdrawals and change of land use for construction purposes. Satellite technology and remote sensing becomes an important source of meteorological data acquisition to complement the traditional methods of meteorological data collection. Natural resource management requires the application of environmental meteorology and GIS into planning and enforcement of laws to set limits for use, consumption and expansion into sustainable Disaster Risk Reduction planning. Use of AI and other technologies has expanded into disaster management activities of prevention and mitigation. Fragile disaster zones have high vulnerability and pose a great risk for human and nonhuman security. Governments around the world have been investing large amounts of money in obtaining and implanting technology for forecasting and disseminating early warnings. In the process, the use of technology demands participation of the community in efforts to save lives and property. Both the Hyogo Framework for Action (2005–2015) and the Sendai Framework (2015–2030) refer to early warning systems (EWSs) as an indispensable tool to save lives during disasters. Nevertheless, the approaches taken to establish EWSs have tended to be fragmented, ad hoc and isolated from communities and local needs.[11] In recent years, laboratory-based EWSs (LBEWSs) have emerged to compensate for this loss. LBEWSs attempt to absorb as much meteorological and geospatial information as the most sophisticated remote sensing and satellite imagery will allow. The latest computational model of CoastalDEM (digital elevation model), in contrast to the previous model, Nasa's Shuttle Radar Topography Mission, developed by Kulp and Strauss (2019), is another turning point in the evolution of disaster management as it exposes the fault lines of adopted projections which help to delay or evade disaster planning. Yet, although these technologies have provided the basis for warning the public of impending disasters, they have often not led to required responses from the administration and institutional systems.

Ethical Governance Out to Turn Tables of Disaster Management The coasts have been ignored despite the increased frequency and impact of disasters cutting through them. Lack of disaster management and repeated institutional failures to redress legal compliance has turned enchanting ballads into a mournful elegy or a funeral hymn as coasts lose their pristine luster. Despite such a tremendous challenge to the survival of coastal life, authorities continue to work in a generic manner rather than through determined strategies which are sustainable, wholesome and empowering. From coastal erosion to overkilling of sea predators, combined with an overload of plastics and concrete constructions, coastal governance has failed to prevent any of this blatant defiance. This gap in disaster management hurts as this relates to a human call for duty beyond a salary's proportional requirement. Disaster management is not a fixed-time office job. It demands the same commitment as a doctor who never sleeps with his phone switched off. Unfortunately, disaster management has been dealt with in a generic manner and only a handful of officials across the country have stepped in with a life commitment. The author's previous paper titled 'Coastal Ballads and Conservation Ironic: Understanding Implementation Slippages of CRZ Law' (Singh 2016), forewarned of a disaster as follows, but until the massive floods of 2018 the state disaster management authority preferred to live in denial:

> …this paper highlights that any wavering on implementing the CRZ law will make areas close to waterbodies vulnerable to disasters amounting to huge socio-economic destruction and loss of lives (Singh 2016, p. 70).

The coasts are one of the most vulnerable areas and need serious political attention and an administrative commitment to preparedness. In the case of the 2018 floods in Kerala, alert signals were already in place with the publication of the Madhav Gadgil Report that stated that 124 villages in the Western Ghats were highly vulnerable to floods and landslides. Yet, Kerala consistently witnessed unbridled land grabbing, expansion of tourism, migration from neighborhood states and construction in fragile eco-zones. When the State Disaster Management Authorities were interviewed during the research team's visit in September 2015, they expressed their strong displeasure over what they saw as the injustices and discrimination of the CRZ Notification of 2011. It was shocking to see that most disaster management, CZMA and municipal administrators were in mission mode to popularize the opinion held by their political executive without even giving a thought to the consequences of their campaign. When the CPM, Member of the Legislative Assembly S. Sharma, moved an adjournment motion, the then Chief Minister Oommen Chandy had already declared that the new regulations would not be implemented. Even Archbishop Francis Kalarakkal of the Kerala Latin Catholic Association had provoked sentiments against these regulations as an existential protest of the fisherfolk.[12] Notwithstanding these regulations, Kerala encouraged and promoted defiance of all CRZ rules of the 2011 Notification. When the 2018 deluge occurred, the consequent disaster exposed a decade-long defiance of coastal management rules.

The government is responsible for many disasters through its soft-peddling attitude on accountability laws. This is evident in the manner in which the present government became sympathetic to Kerala's (as well as Maharashtra's and Karnataka's) objections to the 2011 CRZ Notification. It pacified the critics and especially the fisherfolk, for whom the concern was an existentialist crisis, by constituting the Shailesh Nayak Committee to re-examine the previous Notification which they found to be too stringent. The new CRZ Rules were approved on December 28, 2018 (Notified on January 18, 2019), and shockingly came immediately after the devastating Kerala disaster. The rules have now been amended to bring increased flexibility in decision making to develop coastal areas as per the demands faced by state governments. The benefits of the changed laws as mentioned in the approved document suggest that 'The proposed CRZ Notification, 2018 will lead to **enhanced activities in the coastal regions** thereby **promoting economic growth** while also **respecting the conservation principles** of coastal regions. It will not only **result in significant employment generation** but also to better life and add value to the economy of India. The new notification is expected to rejuvenate the coastal areas while reducing their vulnerabilities'. Therefore the lessons of the 2018 deluge remain unlearned as economics is given priority over ecology without creating parameters of ecological economics to make the builders and tourist companies better regulated. The following changes have been implemented in the amended document which now decides the future of the coasts:

1. The 2011 Notification to keep the Floor Space Index or Floor Area Ratio for CRZ-II (Urban) areas frozen as per 1991 Development Control Regulation levels was opened to allow for construction projects and to enable redevelopment of these areas to meet the emerging needs.
2. The No Development Zones (NDZ) of CRZ III areas stand reduced from 200 m to 50 m in densely populated rural areas but for others with a population of less than 2161 per square kms the NDZ remains at 200 m.
3. Construction for tourism purposes is allowed in CRZ-III areas with a minimum distance from a High Tide Zone to be limited to 10 m.
4. The process of granting clearances to projects has been simplified or relaxed in the rush to streamline them. Only such projects/activities which are located in the CRZ-I (Ecologically Sensitive Areas) and CRZ-IV (area covered between Low Tide Line and 12 nautical miles seaward) will be dealt with for CRZ clearance by the Ministry of Environment, Forest and Climate Change. The powers for clearances with respect to CRZ-II and III have been delegated to the states.

Judicial Intervention Brings Hope When Administrators Escape Accountability. The above changes in CRZ laws have been brought about by the recommendations of Dr. Shailesh Nayak Committee to Review CRZ Notification. The suggestions reflect a typical cost-benefit assessment problem of science pedagogy due to which a fragmented framework is revealed that is unsuitable for disaster assessment. In a world where disasters have become a frequent occurrence a more coordinated framework of rules is expected for protecting the coasts and for meaningful coastal mapping. However, what one gleans is a silo of 'development' which fails to achieve this development in a symbiotic manner so that vulnerability is reduced. The Notification makes no mention of the powerful role which the Disaster Management Authority at the Centre and State level should play in mapping as per geospatial and vulnerability concerns. It is only at a later point, section 5.4 of CRZ-IV, that the Notification mentions the '*the concerned State Government which shall submit the draft Environmental Impact Assessment report (EIA) with Environmental Management Plan (EMP) and the draft Risk Assessment Report with Disaster Management Plan (DMP) …. in accordance with the procedure laid down under the Environment Impact Assessment (EIA) notification number S.O. 1533(E), dated the 14th September, 2006*'.

Many obligations which the governance framework of the Disaster Management Act 2005 assigns to local administrators are not combined to provide synergy and direction, which could have provided a better guarantee against disasters. A lack of clarity in terminology used, such as 'strategic purposes', 'public utilities' and 'eco-tourism', only created larger implementation risks by obfuscating the objective of protection of coasts. Instead of clearly defining these terms it has been left to state governments to find meanings for them. The crucially significant roles of the National and State disaster management authorities in providing a better coordinated, holistic strategy are ignored to the extent that the new laws would drag business as usual at the coasts to the jubilation of realtors, tourist industry and other commercial activities. Without assigning a lead role to disaster management authorities the worst casualty would be the coastal and marine ESAs, mangroves, beaches, coral reefs and so on in controlling coastal erosion, shoreline change and saltwater intrusion and serving as a natural defense against coastal hazards such as storm surges, cyclones and tsunamis (as given in Annexure 1, pp. 43–45).

Consequently, the Hon'ble Supreme Court has come down heavily on the Kerala government as well as the builders and municipal corporation for their defiance of the CRZ.[13] Justice Arun Mishra and Justice S. Ravindra Bhat added certain comments and observations in their judgment that reveal that the judiciary has grown weary of the non-compliance and defiance of law that are becoming a norm with governments. The observations made in the judgment highlight the intentional and opportunistic excuses made to escape the law and accountability in order to allow the continuity of business as usual. The following points picked out from the historic bold judgment may set an example for other courts to follow in achieving accountability on coastal degradation and destruction through legal defiance:

(a) directed freezing of assets of builders and promoters who were involved in the construction of illegal buildings in the coastal zone areas of Kochi;
(b) asking government to recover the interim compensation amount from the builders and promoters;
(c) reprimanding the Kerala Chief Secretary for delays and then asking him to stay away from the hearing;
(d) establishing that it is the court's primary concern to ensure no construction in the eco-fragile coastal zone.

The Court order, which is boldly worded and also exposes the political intentions of administrators, will go down in the history of disaster management:

> *It is due to such violation and tortuous action, the entire environment is being degraded and coastal zones are being illegally occupied. It appears that the authorities, rather than preventing the violations, are trying to mobilize the public opinion and time has come to hold them responsible for their active connivance in such activities of degrading the environment and violation of the coastal zone regulations etc. (p. 1)*

Conclusion

The law of conservation is not inherently weak. However, what makes it weak is the combination of the failure of state leadership, administrative and judicial corruption and ecological awareness. There are many factors which prevent sensitization, resonance and rigor in environmental governance. The most important one is the need for public feedback (and the application of the *audi alteram partem rule to expect all parties to be heard appropriately before taking a decision*) and a regular supervisory and evaluation committee which can best be arranged through the central ministry. In the Case of protecting CRZ of Kerala Backwaters Justice A.V. Ramakrishna Pillai observed,

> *The purpose of these laws is to preserve nature for posterity. If the violation of the laws is allowed to become the order of the day, the existence of life would be at peril. Right to life guaranteed by the Constitution takes innumerable rights, including the right to enjoy nature in the present form. Indiscriminate invasion of nature to the detriment of others is an invasion of right to life.*

All this could be achieved if public institutions perform as custodians of the trust property rather than agents of its destruction and a cause for disaster. The lesson of this chapter is that disaster management can be improved if urban planning is merged with environmental resource planning under the leadership of ethically conscientious administrators. Disaster management is not a department to be set up for material gain but rather to seek meaning in philanthropic activity that helps life on the planet. Fortunate are those administrators who evolve by being part of this mission.

Notes

1. Kankara et al. (2018).
2. Bhattacharya, Bibek (2019) mentions a study by Anjal Prakash, associate professor at The Energy and Resources Institute (Teri) School of Advanced Studies, who observes, 'The sea level will continue to rise. It is projected to reach around 30–60 cm by 2100 even if the greenhouse gas emissions are sharply reduced and global warming is limited to well below 2 degrees Celsius'.
3. 'India Installed Capacity' (PDF). Retrieved October 8, 2018.
4. Fackler (2012) and Yamaguchi (2012) found the disaster to be avoidable and a result of company lapses.
5. Singh (2018) and the SCDR-JNU research team traveled through the flood-affected rural rim. The project was funded by the university's UPE II grant.
6. 'Deltas Sinking and Shrinking as Dams Curb Steady Flow of Fresh Water to the Coasts', *The New Indian Express*, January 28, 2015, 09:09 pm, Study of Coastal Zones National Centre for Sustainable Coastal Management, '*India has 7500 km coastline (~5400 km on the mainland) and about 250 million people live within 50 km distance from the shore. India's mainland coastal stretch of 5422 km have undergone tremendous changes due to varying natural and human induced coastal activities*' (http://ncscm.org/cms/geo/pdf/research/High%20resolution%20Erosion.pd) (SANDARP 2014).

7. Costanza et al.'s (1997) valuation of the world's ecosystem services and natural capital, May 15, p. 253.
8. Ajayan in Livemint on October 8, 2007 writes, 'The Kerala government's push for an ordinance to bring plantations under the purview of the Ecologically Fragile Land (EFL) Act has rattled the plantation sector in the state'.
9. By using the power of robotics and artificial intelligence, Robin Murphy is helping people rebuild their lives (#womeninstem).
10. Article 141 provides that the law declared by the Supreme Court shall be binding on all Courts within the territory of India. It provides that in order to carry out complete justice, the Supreme Court will have the power to pass any judgment, decree or order as is necessary.
11. South Asia Disasters (2018) establishes the status of Early Warning System to communities.
12. Read more at: http://timesofindia.indiatimes.com/articleshow/29918143.cms?utm_source
13. The Kerala State Coastal Zone Management Authority vs. Maradu Municipality & Ors. MA No. 1808–1809 of 2019 in Civil Appeal Nos. 4784–4785, 4786–4789, 4790–4793 of September 23, 2019.

References

Ajayan. (2007). Kerala Move to Amend Key Act Relating to Plantations Draws Flak. *Live Mint*, 8th Oct, 12:16 AM. Retrieved 20th Sept 2019.

Bhattacharya, B. (2019). Mumbai and Kolkata Face the Wrath of Sea Level Rise. *Live Mint*, 1st November, 12:24.

Costanza, R., d'Arge, R., de Groot, R., Farberk, S., Grasso, M., Hannon, B., Limburg, K., Naeem, S., O'Neill, R., Paruelo, J., Raskin, R. G., Sutton, P., & Belt, M. v. d. (1997). The Value of World's Ecosystem Services and Natural Capital. *Nature, 387*, 253.

Deltas Sinking and Shrinking as Dams Curb Steady Flow of Fresh Water to the Coasts. *The New Indian Express*, 28 Jan. 2015, 09:09 pm, Study of Coastal Zones National Centre for Sustainable Coastal Management.

Fackler, M. (2012). Japan's Power Company Admits failings on Plant Precautions. *The New York Times*, 12th October. Retrieved 16th Sept 2019.

Kankara, R. S., Ramana Murthy, M. V., & Rajeevan, M. (2018). *National Assessment of Shoreline Changes Along Indian Coasts- A Status Report for 26 Years 1990–2016*. Chennai: National Centre for Coastal Research, Ministry of Earth Sciences.

Kulp, S. A., & Strauss, B. H. (2019). New Elevation Data Triple Estimates of Global Vulnerability to Sea-Level Rise and Coastal Flooding. *Nature Communications, 10*, 4844. https://doi.org/10.1038/s41467-019-12808-z.

SANDARP News Bulletin. (2014, May 7). *Sinking and Shrinking Deltas: Major Role of Dams in Abetting Delta Subsidence and Effective Sea Level Rise*. South Asia Network of Dams, Rivers and People.
Singh, A. (2016). Coastal Ballads and Conservation Ironic, Understanding Implementation Slippages of CRZ Laws. *Economic and Political Weekly, LI*(7).
Singh, et al. (2018). *2018 Kerala Floods, Report on Governance and Legal Compliance*. Special Centre for Disaster Research Publications, JNU.
South Asia Disasters. (2018). Early Warning System (EWS) and Community Resilience to Floods. *Asian Early Warning Systems: A View*. Special Issue, No. 170, p. 8. Available at southasiadisasters.net
Yamaguchi, Mari (12 October 2012). *Japan utility agrees nuclear crisis was avoidable. Boston.com. Associated Press*. Retrieved 13th Oct 2012.

Amita Singh is Professor of administrative reforms at the Centre for the Study of Law and Governance, JNU, New Delhi, and founder chairperson and transdisciplinary scholar at the Special Centre for Disaster Research, JNU, New Delhi.

The manufacturer's authorised representative in the EU is Springer
Nature Customer Service Centre GmbH, Europaplatz 3, 69115 Heidelberg,
Germany. If you have any concerns regarding our products, please
contact ProductSafety@springernature.com

Printed and bound by CPI Group (UK) Ltd, Croydon, CR0 4YY
25/09/2026
02222903-0003